Topics in Mathematics

Ideas for the Secondary Classroom

BELL & HYMAN
London

First published in 1979 by
BELL & HYMAN LIMITED
Denmark House
37–39 Queen Elizabeth Street
London SE1 2QB

ISBN 0 7135 1999 1

Printed in Great Britain by
Robert MacLehose & Co Ltd
Glasgow

Preface

It has been suggested that there is a great deal of useful material hidden away in Classroom and Mathematical Notes published in the *Mathematical Gazette*: material which may have been forgotten but which could well be a source of ideas for teachers today.

The working party set up to make a selection of these Notes has chosen the best of the available material published during a period of about forty years. In nearly all instances the units and notation have been updated. In some cases notes have been combined under a single idea and it was found necessary to adapt others for the sake of clarity and conformity to current practice.

The selection of the chapter headings is quite arbitrary but it is hoped that these divisions will prove appropriate and helpful. At the end of each chapter will be found a list of references to the Notes which have been used as a basis for this book. At the end will be found an index which refers to topics whose location is not immediately obvious from the section headings.

MEMBERS OF THE WORKING PARTY

Mr P. BAILEY
Dr D. J. BAYLIS
Dr J. BIBBY
Mr J. M. BRADLEY
Dr V. BRYANT
Mr M. J. CAHILL
Mr D. C. CARTER *Editor*
Mrs B. M. DENT
Mr C. J. GRIFFITHS
Mr D. M. HALLOWES *Chairman*
Mr R. KAPADIA
Mr D. R. KIDDLE
Mr D. G. MACLEOD
Mr P. J. READ
Mr P. T. ROBBINS
Mr I. STEGGLES
Mr D. STRUTHERS
Mr D. UMBERS
Mr M. F. WORBOYS

Contents

1 ARITHMETIC
1.1 Some notes on elementary arithmetic *page* 11
1.2 Egyptian mathematics 14
1.3 Approximations to square roots 17
1.4 Using Pythagorean number-triples 19
1.5 Diagrams to illustrate the geometric series 20
1.6 1961 23

2 PROBABILITY and COMBINATORICS
2.1 Beating the bookies 25
2.2 Problem arising from the crap game 26
2.3 Squash chances 28
2.4 Football probabilities 29
2.5 Runs of successes and failures 31
2.6 Permutation problems 31
2.7 Problems of paths 36
2.8 Transport problems 39
2.9 How unlikely? 41
2.10 The hat check problem 42
2.11 A broken stick 44
2.12 A case of local probability 46
2.13 Randomness 47

3 ALGEBRA
3.1 Solution of quadratic equations by geometrical constructions 50
3.2 A graphical representation of quadratic equations 52
3.3 A nomogram for the solution of quadratic equations 53
3.4 An interesting equation 56
3.5 The condition for the cubic equation to have real roots 56
3.6 The real roots of a cubic equation 58
3.7 Complex roots of a cubic equation from a graph 58
3.8 Graphs of cubic functions 60
3.9 The Remainder Theorem 62

3.10 The inequality between the arithmetic and geometric means 65
3.11 Applications of the inequality of the means 67
3.12 The Binomial Theorem for any index 67
3.13 Pascal's Triangle 68
3.14 Sums of powers of the natural numbers; proofs by dissection methods 71
3.15 Sums of powers of the natural numbers; proofs by algebraic methods 76
3.16 A single-scale nomogram 80
3.17 The sign of a rational function 81
3.18 Inverting a formula 81
3.19 Multiplication of triangular matrices 82
3.20 Illustrations for $x^0 = 1$ 83
3.21 An example on mathematical induction 83
3.22 The mixture as before 85
3.23 Crossing the desert 85

4 CALCULUS and ANALYSIS
4.1 An area property of the curve $y = x^n$ 89
4.2 Extending the trapezial rule for the area 89
4.3 Functions, differences and the fundamental theorem of the calculus 90
4.4 How arbitrary can a constant of integration be? 93
4.5 Integration of inverse functions 94
4.6 Some integration techniques 95
4.7 An old problem: a donkey tethered to the edge of a circular field 97
4.8 Investigating stationary points 98
4.9 The gradient of the sine curve 99
4.10 Differentiation of logarithms 101
4.11 Another approach to the logarithmic and exponential function 102
4.12 Continuous interest and the exponential function 103
4.13 Limit of $\sin\theta/\theta$ as $\theta \to 0$ 104
4.14 The convergence of x^n 104
4.15 The harmonic series 105
4.16 A topological approach to continuity 105
4.17 Mean value theorems for higher derivatives 108
4.18 Reciprocals as an introduction to iteration 109

4.19 Some irrational number approximations 109
4.20 Miscellaneous methods of finding approximate values of logarithms 112
4.21 Radius of curvature in pedal co-ordinates 113
4.22 A source of error in solving differential equations 114
4.23 Connecting the forms of the solutions of differential equations 115

5 GEOMETRY
5.1 Directed doodles 117
5.2 Pythagoras' Theorem 118
5.3 A curious triangle 119
5.4 Converses 119
5.5 Integral-sided geometric figures 120
5.6 Rectangling the circles 123
5.7 The maltitudes of a cyclic quadrilateral 124
5.8 The 'alternate segment' theorem 125
5.9 Mathematical tiles 128
5.10 Constructing an angle of 30° 129
5.11 Uses of a geometric puzzle 130
5.12 The perpendicular from a point to a line 132
5.13 To construct a triangle given the angle A, and the sums of sides $c + a, b + a$ 133
5.14 Tangrams and incommensurables 133
5.15 From the pentagon to the dodecahedron 134
5.16 A well-known geometrical problem 135
5.17 When are triangles congruent? 138

6 TRIGONOMETRY
6.1 Angles whose sum is $\pi/4$ 141
6.2 The sine and cosine of obtuse angles 142
6.3 To determine the height of an inaccessible object 144
6.4 The Cosine Rule 145
6.5 Model to illustrate latitude and longitude 146
6.6 Great circles on a sphere 148
6.7 A model to demonstrate the variation of $\sin x$ and $\cos x$ 149
6.8 The addition formulae 150

7 VECTORS
7.1 Use of vectors for finding OH^2, OI^2 in a triangle 157
7.2 The triple vector product formula 158

8 MECHANICS
8.1 Geometrical proof of $v^2 = u^2 + 2fs$ 163
8.2 A first lesson on kinematics—Application of graphs 164
8.3 A walk in the rain 165
8.4 Implicit assumptions in standard problems 167
8.5 Some remarks on the coefficient of restitution 169
8.6 An elementary collision problem 172
8.7 Theorems on centroids 173
8.8 Further centre of gravity problems 174
8.9 Moments of Inertia 176
8.10 Proof of the formula for acceleration of a point moving with uniform speed in the circumference of a circle 177
8.11 The elementary theorems in analytical statics 178
8.12 The reduction of the mass-accelerations and the momentum of a rigid lamina 179
8.13 A geometric side-line on 8.12 182
8.14 A problem on friction 182
8.15 Braking efficiency in motor cars 184
8.16 Note on a class of problems in elementary dynamics 185
8.17 Problems on variable mass 187
8.18 Motion of a rocket 189

INDEX 192

1 Arithmetic

1.1 Some Notes on Elementary Arithmetic

I It sometimes happens that a class has not properly grasped something which should have been done some time ago and one wishes one could use some form of decimal apparatus, but it is not available. In such a case, refer to the units as single *sticks* of firewood, the tens as *bundles* of ten, the hundreds as *sacks* of ten bundles and, if required, the thousands as *carts* of ten sacks.

II Many children seem lost when they first meet a problem such as to divide a certain sum between father and son so that the father has three times as much as the son. This unequal division puzzles them. It can be converted into a problem of equal division by supposing that the father has three pockets and the poor little son has but one, then the problem is to divide the money between four pockets.

III If children are taught the easy Roman Numbers (i.e. M.D. C.L. X and I) they not only have an alternative numbering system extremely useful for sub-paragraphs in note-taking in any subject at any standard, but they also seem to get a better grasp of our Arabic number system, probably because when they see the symbol '4' for example; they realise that it is FOUR of something, either MMMM or CCCC or XXXX or IIII.

IV Quite a number of problems, both in real life and in examination papers, are really only disguised cases of what pupils at one school used to call the 'Concert seat problem'. An easy example is: at a certain charity concert at a private house, there were 24 seats, some at 50p and some at 30p. If the total money received was 880p how many of each kind were there? This CAN be done by algebra, but surely the easiest method is to suppose that all the seats were 30p but that some had cushions which, clearly, were 20p each. Then the 24 seats at 30p come to 720p leaving 160p for cushions, and as these were 20p each, there were 8 cushions: i.e. 8 of the more expensive seats and the rest the cheaper ones.

V It is sad how many children really do not realise that, for example, two fifths is the same as a fifth of two. One of the best ways of showing this is by drawings of round, flat cakes. The left figure shows two fifths in the usual way while the right figure represents one cake on top of the other. Then, in the right figure, one takes one fifth of the two cakes and the resulting two slices together are clearly the same as the double sized slice in the left figure.

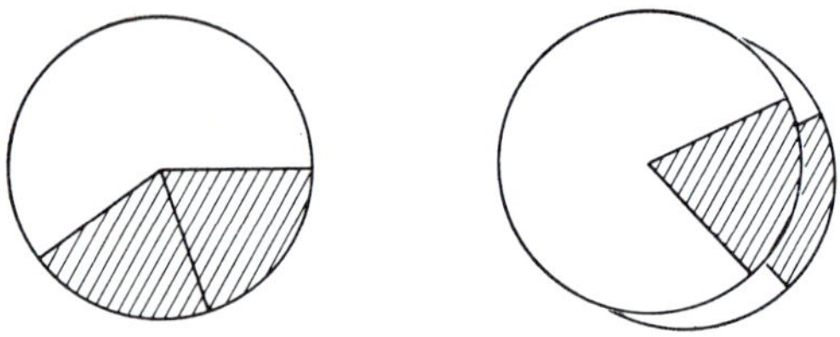

VI 'Seeing is believing', and a visual aid certainly does help some pupils, even if not all. This diagram may be useful in supplementing whatever method is usually used to deal with this point.

Take any three unequal lengths *a*, *b*, *c*, and the rest follows from an examination of the diagram.

Naturally the diagram without a word of explanation from the teacher would not be of much use; but it can be seen that, in the last line, the *c* has, as it were, re-appeared. By inspection

$$a-(b-c)=a-b+c.$$

This may well be followed either by each member of the class choosing his own numerical example and working it out by two different routes, for example,

$$10-(7-2) = 10-5=5,$$

or

$$\begin{aligned} 10-(7-2) &= 10-7+2 \\ &= 12-7=5, \end{aligned}$$

or if preferred one could show diagrammatically that

$$\begin{aligned} a-2(b-c) &= a-2b+2c, \\ a-(b-2c) &= a-b+2c, \\ a-2(b-2c) &= a-2b+4c. \end{aligned}$$

These too can be followed by numerical examples.

The following mnemonic amuses older pupils—particularly if it is mentioned that it has actually been given seriously as a teaching method. If we wrote two minus signs in succession, thus – – , they might get run into each other and mistaken for just one — . So we turn one on end and write them like this: +.

VII A motorist travelled 100 km at 35 km/h and returned at 25 km/h. What was his average speed for the 200 km journey? Inevitably one receives from some students the answer 30 km/h. That this answer is wrong, may be illustrated by considering the same problem but with speeds of 35 km/h and 25 km/h being replaced by 100 km/h and 0 km/h respectively. Clearly in this case the motorist never returns, which offends the suggestion that his average speed was 50 km/h.

If the outward journey had been at 1 km/h there would have been no finite speed at which the motorist may have returned to make his average speed 2 km/h. It seems therefore that his average speed is bounded above by a function of the slower speed. This hypothesis prompted an investigation of the following more general problem.

A racing motorist completed n laps of a course at average lap speeds of $v_1, v_2, v_3, \ldots, v_n$. Is there an upper bound of the average speed, V, for the n laps, which is a function of the least average lap speed v_i?

In fact $v_i \leqq V < nv_i$.

Proof

(i) $v_i \leqq V$ by hypothesis.

$$\text{(ii)}\ V = \frac{\text{total distance}}{\text{total time}} = \frac{nd}{d\sum_{j=1}^{n}\frac{1}{v_j}} = \frac{nv_i}{\sum_{j=1}^{n}\frac{v_i}{v_j}},$$

where d is the distance of one lap.

But $\sum_{j=1}^{n}\frac{v_i}{v_j}$ is a series of positive terms containing $\frac{v_i}{v_i}$ (= 1).

Therefore $\sum_{j=1}^{n}\frac{v_i}{v_j} > 1$, and $\dfrac{nv_i}{\sum_{j=1}^{n}\frac{v_i}{v_j}} < nv_i$, since nv_i is positive.

i.e. $V < nv_i$.

Intuitively one feels that if there is an upper bound of the average speed which is a function of the least lap speed, there ought to be a lower bound of the average speed which is a function of the greatest lap speed. This however, is not the case, for if $(n-1)$ of the lap speeds are prescribed, a remaining lap speed may be determined sufficiently small to make the average speed as small as we please.

VIII It may perhaps be helpful to point out that on a scale in which the radius of the earth's orbit round the sun is represented by one inch, the light-year is very closely represented by one mile; for,

$$\frac{186\,000 \times 60 \times 60 \times 24 \times 365}{93\,000\,000 \times 1760 \times 36} = \frac{219}{220}.$$

1.2 Egyptian Mathematics

In the Rhind papyrus the area of a circle is found by reducing the diameter by $\frac{1}{9}$ and squaring the result, i.e. $(d-d/9)^2$ in modern terms. In speculating as to how the ancient Egyptian arrived at this rule one recalls that he worked with *unit* fractions only, and that his ruler reflected this system, the successive units of the ruler being marked in halves, thirds, fourths, etc. The ruler was, therefore, well adapted for dividing a line into any number of equal parts. To discover the above rule, did the Egyptian reduce the diameter by various unit fractions in turn, construct a square on the reduced diameter, and compare area of square and circle in each case? (We know that he drew on a squared surface, and that his units of area were based on a square. So he could compare areas by counting squares.)

This method was tried, drawing on 1 cm squared paper, quadrants of radius 10, 9, 8, and squares of sides 9, 8, 7, and comparing areas by counting grid squares. One concludes that the Egyptian could have discovered, without much difficulty, that circle 9 and square 8 were most nearly equal; and $\pi x^2 - 4(x-1)^2$, for $x = 10, 9, 8$, can be checked by calculation. By solving $\pi d^2 = 4(d - xd)^2$, one finds that $(d - 9d/80)^2$ is a nearer approximation than $(d - d/9)^2$. Further, since $\frac{9}{80}$ is reducible to unit fractions, as $\frac{1}{8} - \frac{1}{80}$ or $\frac{1}{10} + \frac{1}{80}$, it would have been quite possible for an Egyptian to have discovered this nearer approximation by an extension of the empiric method being tested.

But this thought provoked an anti-climax. Was not this empiric method far too modern for an ancient Egyptian? Was it not more likely that his rule came from a sudden inspiration? Another line was tried. By drawing circles on a squared surface the Egyptian would no doubt learn which points of intersection of the grid lines lay on particular circles. He certainly knew, about 5000 years ago, that (3, 4) lies on a circle of radius 5 units. Has the 9-unit circle any special feature? Did drawing it give him a clue? Using Pythagoras, $9^2 - 4^2 = 65$, or 8^2 approx. (the approximation is near enough for ancient Egyptian drawing). So (4, 8) lies on a 9-unit circle.

The solution is immediately obvious. *A square, and a circle through the $\frac{1}{4}$-points of its sides, have approximately the same area.* Here was the origin of the ancient Egyptian rule! But to establish this as a fact needs further evidence.

Taking side of square $= 16$, then A (area) $= 256$. $R^2 = 4^2 + 8^2 = 80$. So $A/R^2 = 3\frac{1}{5}$.

Now certain measurements of the Great Pyramid are held by some authorities to show that the Egyptians knew how to calculate the circumference of a circle from the diameter. There is, however, no direct evidence of this knowledge. So the discovery of the $\frac{1}{4}$-point circle's significance sets one wondering whether there was any other-point circle which would give a simple rule for finding the circumference. Taking side of square $= 2$, and circle of equal perimeter passing through $(x, 1)$, then

$$2\pi\sqrt{1 + x^2} = 8.$$

So $x = \cdot 787$. $x = \frac{3}{4}$ is the nearest approximation which gives $R(=\frac{3}{4})$ exactly measurable in the same units ($\frac{1}{4}$) as x. This circle passes through the $\frac{1}{8}$-points of the sides of the square, and C (perimeter of square or circle) $\div d = 3\frac{1}{5}$, as for A/R^2 above. This suggests a method of approaching $C/d = A/R^2 = \pi$ in school, based on the assumption that a square has approximately the same area as the $\frac{1}{4}$-point circle, and the same perimeter as the $\frac{1}{8}$-point circle. Then $C = 4(d - d/5)$. ($x = \frac{4}{5}$ gives the $\frac{1}{10}$-point circle, but no unit fraction formula for C.)

It will be observed that the $\frac{1}{8}$-point circle is a 5-unit circle, each $\frac{1}{8}$-point being a corner of a 3-4-5 triangle. There is nothing to suggest that the Egyptians connected this circle with a square.

The ancient Egyptians, as is known from the Rhind papyrus, had no notation for fractions with other than a unit numerator. Other fractions were represented by the sum of fractions with unit numerator. How did they come to use such an apparently obscure system?

Some primary school children were given a number of identical paper rectangles, and each of them was told between how many people their rectangles were to be accurately shared.

A number of solutions were obtained by laying the rectangles end to end and dividing this larger rectangle into the required number of parts, like this:

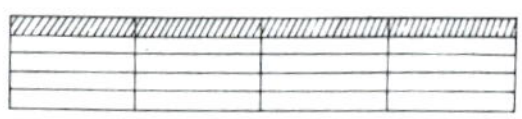

(4÷5)

Others solved the problem after this manner, though usually not at all accurately:

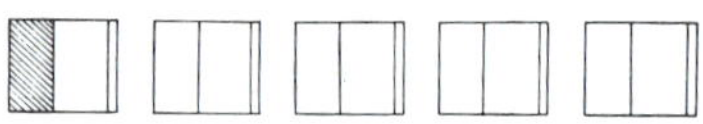

(5÷11)

However, a few pupils, with less foresight than the others, but no lack of confidence, produced quite different solutions, as in these two examples:

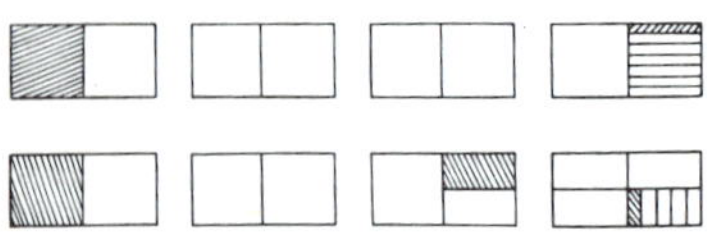

These examples are equivalent to $\frac{4}{7} = \frac{1}{2} + \frac{1}{14}$ and $\frac{4}{5} = \frac{1}{2} + \frac{1}{4} + \frac{1}{20}$.

Might this simple process, sharing part of the whole, then sharing out the remainder, be the method by which the Egyptians obtained their results? It appears an ugly and even absurd system to us, but we are used to performing complicated operations with fractions. If the Egyptians needed no more than to be able to compare fractions in magnitude, or a few other simple operations, they may have found $\frac{1}{2}+\frac{1}{14}$ and $\frac{1}{2}+\frac{1}{8}$ easier to manage than their modern equivalents $\frac{4}{7}$ and $\frac{5}{8}$, and because their system was sufficient for them, they were never impelled to improve on it.

1.3 Approximations to Square Roots

I To find $\sqrt{2}$. The nearest integer to $\sqrt{2}$ is 1, so $\sqrt{2}-1 < \frac{1}{2}$.
By repeated squaring we have

$$-2\sqrt{2}+3 < \tfrac{1}{4},\ -12\sqrt{2}+17 < \tfrac{1}{16},\ -408\sqrt{2}+577 < \tfrac{1}{256}, \ldots$$

From the last relation we may with negligible error write

$$-\sqrt{2}+\tfrac{577}{408} = 0, \quad \text{or} \quad \sqrt{2} = \tfrac{577}{408}.$$

Solutions of the Pellian equation are given by the numerator and denominator at any stage.

To find $\sqrt{7}$, for example, one starts with $3-\sqrt{7}$ and it can easily be shown that it is $\frac{127}{48}$ very nearly, and that $x^2-7y^2 = 1$ has solutions $x = 8, y = 3$ and $x = 127, y = 48$.

One is able to consider Fermat's two theorems (the one via solutions of $x^2+y^2 = z^2$, the other via the decimals of $\frac{1}{7}, \frac{1}{17}, \frac{1}{19}$), the relation between the first and second halves of the recurring groups in these decimals, the peculiarity of $2^{25}+1$, etc.

It is of course just as easy to see how a cube root is found, or for that matter how a cubic is solved. For instance $\sqrt[3]{2}-1 < \frac{1}{2}$, so, writing x for the root,

$$x^2-2x+1 < \tfrac{1}{4},\ x^4-4x^3+6x^2-4x+1 = 6x^2-2x+7 = \tfrac{1}{16},$$
$$36x^4-24x^3-80x^2+28x+49 = -80x^2+100x+1 > \tfrac{1}{256}, \ldots$$

Convergents to the root are found by solving $p_nx^2+q_nx+r = 0$. The analogy with the convergents of a continued fraction is brought out by regarding the latter as solutions of $q_nx-p_n = 0$.

Given an equation of any degree with at least one real root, we can in this way form an equation whose degree is one less, and one root of which is very close to a root of the given equation. But no one would assert that for equations higher than cubics this is a good method.

II If r is an approximation to the square root $r+h$ of a, then we can obtain a closer approximation, by using the identity

$$(r+h-r)^2 = (r+h)^2 - 2r(r+h) + r^2$$

and using the fact that h is small compared with r. Neglecting h^2 and replacing $(r+h)^2$ by its value a, we find

$$0 = r^2 - 2r(r+h) + a$$

whence

$$r+h = \frac{r^2+a}{2r}, \tag{1}$$

which is, of course, the expression obtained by applying the usual Newton approximation.

We can, however, go further with this method of derivation, for

$$(r+h-r)^3 = (r+h)^3 - 3(r+h)^2 r + 3(r+h)r^2 - r^3.$$

Putting $(r+h)^2 = a$, we have

$$0 = r^3 - 3r^2(r+h) + 3ra - (r+h)a$$

whence

$$r+h = r\left(\frac{3a+r^2}{a+3r^2}\right), \tag{2}$$

which is a better approximation than (1).

It is possible to go a stage further still along the same lines, treating in the same manner the expansion of $(r+h-r)^4$ from which is obtained the formula

$$r+h = \frac{a^2 + r^2(6a+r^2)}{4r(a+r^2)}, \tag{3}$$

but this is identical with two applications of (1).

An example will show the accuracy of approximations from these three formulae, when used to determine $\sqrt{10} = 3{\cdot}162278$, with $r = 3$, say. Formula (1) gives $r+h = \frac{19}{6} = 3{\cdot}16667$ and the next approximation is $\frac{721}{228} = 3{\cdot}162281$. Formula (2) gives $r+h = \frac{117}{37} = 3{\cdot}162162$ and formula (3) gives $\frac{721}{228}$ again, as previously observed. For practical purposes, recurrence formula (2) gives a close approximation with little computation.

1.4 Using Pythagorean Number-Triples

A very interesting lesson on generalisation, formulae and the use of brackets can be developed from the starting-point of a few well-known Pythagorean number-triples. In each of the following series the first three sets of triples can be written on the board and the class encouraged to continue by analogy.

The first series has successive odd numbers for the side a.

n	a	b	c	$(a+b+c)$	factors	p	q
1	3	4	5	12	3×4	2	1
2	5	12	13	30	5×6	3	2
3	7	24	25	56	7×8	4	3
4	9	40	41	90	9×10	5	4
5	11	60	61	132	11×12	6	5

The following formulae can be guessed, written down, and the various relationships between them proved:

$$a = 2n+1 \quad b = 4 \times (\text{series of triangular numbers } 1, 3, 6, 10\ldots)$$
$$= 2n(n+1).$$
$$c = b+1 = 2n(n+1)+1; \; a+b+c = 2(n+1)(2n+1).$$
$$a^2 = b+c; \; b = n(a+1) = (n+1)(a-1).$$
$$p = n+1, \; q = n; \; a = p^2-q^2, \; b = 2pq, \; c = p^2+q^2.$$

In the second series we take successive multiples of 4 for a.

n	a	b	c	$(a+b+c)$	factors	p	q
1	4	3	5	12	$2 \times 2 \times 3$	2	1
2	8	15	17	40	$2 \times 4 \times 5$	4	1
3	12	35	37	84	$2 \times 6 \times 7$	6	1
4	16	63	65	144	$2 \times 8 \times 9$	8	1
5	20	99	101	220	$2 \times 10 \times 11$	10	1

In this case

$$a = 4n, \; b = (2n-1)(2n+1) = 4n^2-1, \; c = b+2 = 4n^2+1;$$
$$a+b+c = 2 \,.\, (2n)(2n+1); \; a^2 = 2(b+c); \; b = na-1,$$
$$c = na+1.$$

Taking the two statements

$$3^2+4^2 = 5^2 \quad \text{and} \quad 10^2+11^2+12^2 = 13^2+14^2$$

a class with a knowledge of quadratic equations can grapple with the problem of finding the 'next' one or two. They turn out to be

$$21^2+22^2+23^2+24^2 = 25^2+26^2+27^2$$

and
$$36^2+37^2+38^2+39^2+40^2 = 41^2+42^2+43^2+44^2.$$

In the process of finding these results the class will, with any luck at all, notice:

(i) that any such sequence will lead to a quadratic
(ii) that there is always a set of solutions symmetrical about zero
(iii) that the constant term in their quadratic is always negative
(iv) that positive integral solutions always exist.

With these results before them they can then predict the next one or two and test them.

A general approach to this problem, viz, the solution of

$$\sum_{0}^{n-1} (a+n)^2 = \sum_{n}^{2n-2} (a+n)^2,$$

would appear to be suitable for the Sixth. It involves $\sum n$ and $\sum n^2$ and leads to the quadratic equation

$$a^2-2(n-1)^2a-(n-1)^2(2n-1) = 0.$$

A simpler solution is found from the equation

$$\sum_{r=0}^{n} (a-r)^2 = \sum_{r=1}^{n} (a+r)^2$$

which leads to the linear equation

$$a = 4\sum n = 2n(n+1).$$

1.5 Diagrams to Illustrate the Geometric Series

I The concept of convergence of infinite geometric series is often illustrated by subdivision of a straight line or rectangle of unit length into successive segments of lengths $\frac{1}{2}$, $\frac{1}{4}$, $\frac{1}{8}$, . . ., units. By repeated subdivision of the residual length, it is shown that the sum to infinity of the series is one unit. This treatment, however, does not readily extend to values of the common ratio other than $r = \frac{1}{2}$. The following method enables any positive value of r to be considered.

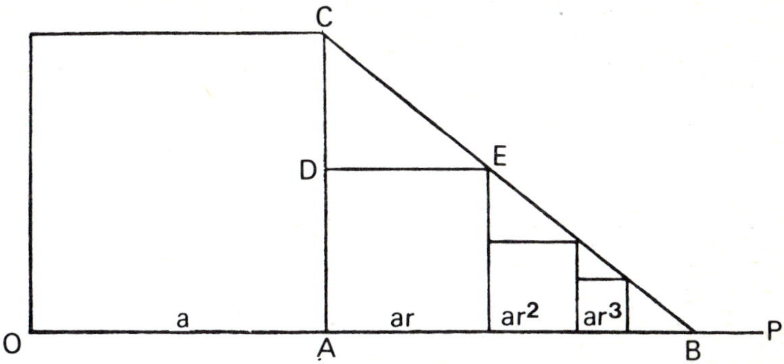

On a base line *OP* (see diagram), mark off adjacent lengths proportional to the numerical values of successive terms of the series, and construct squares on each of these line segments as bases. Three or four squares are usually sufficient. The top right-hand corners of the squares all lie on the straight line *CB*, and therefore, no matter how many more terms of the series are taken, the process cannot be extended beyond the point *B*; *OB* therefore represents the sum to infinity. Taking first the case where $r = \frac{1}{2}$, it is easily shown that $AB = AC = \frac{1}{2}$ unit, hence $OB = 1$ unit.

The method can now be generalised for any positive value of *r*. The corners of the squares always lie on a straight line which cuts the base line at a finite point *B* only when $0 \leqq r < 1$. The gradient of the line *CB* gives a visual indication of the rate of convergence, and measurement of *OB* provides a simple graphical method of obtaining an approximate value for the sum to infinity.

The formula for the sum to infinity can be derived using the same construction. Representing successive terms of the series $a, ar, ar^2, \ldots$ by the bases of successive squares, we obtain, from similar triangles *ABC*, *DEC*:

$$AB = \frac{AC \,.\, DE}{DC} = \frac{a \,.\, ar}{a - ar} = \frac{ar}{1-r},$$

and

$$S_\infty = OB = OA + AB = a + \frac{ar}{1-r} = \frac{a}{1-r}.$$

The formula for the sum of *n* terms of any geometric series with a positive common ratio can be similarly derived.

II Suppose $r < 1$. On OX take $OP = r$, and $OQ = 1$. Draw ordinates PL and QM. Take any point A on QM, and join OA cutting PL at B. Draw BA_1 parallel to OX, and let $AA_1 = a$. The successive terms of the series, $a, ar, ar^2, \ldots$ may then be represented as follows.

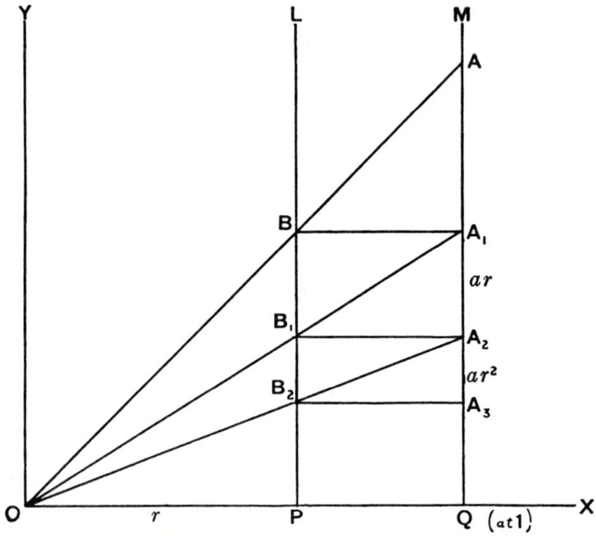

Join OA_1 cutting PL at B_1, and draw B_1A_2 parallel to OX. Then $A_1A_2 = ar$.

Join OA_2, cutting PL at B_2, and draw B_2A_3 parallel to OX. Then $A_2A_3 = ar^2$.

The construction can now be repeated to represent ar^3, ar^4 .... The similarity of the triangles formed may be used to show that the intercepts AA_1, A_1A_2, A_2A_3 represent the terms of the geometric series.

It can be seen that AQ is the sum to infinity, and since

$$AQ/OQ = AA_1/A_1B,$$

we have that the sum to infinity is $a/(1-r)$. The sum to n terms may also be derived.

The diagram may be compared with one formed by taking $r = OP > 1$.

1.6 1961

The year 1961 will be unique in our lifetime for, of the years of this century, only 1961 is unaltered by being written upside down. This last happened in 1881 and will not occur again till 6009. How rare is such an event? Let us call a number *invertible* if it is a positive integer which is unaltered by being written upside down. Then the above question can be conveniently restated as: how many numbers less than 10^k are invertible?

There are 2 invertible numbers less than 10 : 1 and 8. An invertible number between 10 and 100 is either an invertible digit repeated or a pair of digits each of which is the other written upside down, so there are 4 such numbers: 11, 69, 88, and 96. The invertible numbers with three digits can be found by putting 0, 1 or 8 in the middle of each of these numbers. There are therefore 12 invertible numbers between 100 and 1,000. Of course, no number of the form 10^n is invertible unless $n = 0$.

To construct all invertible numbers with $n+2$ digits, where $n > 1$ is even, it is only necessary to insert successively in the middle of all invertible numbers with n digits all invertible numbers with 2 digits and 00. Similarly, to construct all invertible numbers with $n+2$ digits where $n > 1$ is odd, we insert successively on either side of the middle digit of each invertible number with n digits, the digits of each invertible number with 2 digits and the digits 00. Therefore, for $n > 1$, there are five times as many invertible numbers with $n+2$ digits as invertible numbers with n digits. Consequently, for $n > 1$, there are $4 \,.\, 5^{(n/2)-1}$ invertible numbers between 10^{n-1} and 10^n if n is even and $12 \,.\, 5^{[(n-1)/2]-1}$ if n is odd.

Let A_k be the number of invertible numbers less than 10^k. Then, by taking the sum of the numbers of invertible numbers with at most k digits, for $m > 0$,

$$A_{2m} = 2+\sum_{n=1}^{m} 4 \,.\, 5^{n-1}+\sum_{n=1}^{m-1} 12 \,.\, 5^{n-1}$$

$$= 2+4 \,.\, 5^{m-1}+16 \,.\, \sum_{n=1}^{m-1} 5^{n-1}$$

$$= 2+4 \,.\, 5^{m-1}+4(5^{m-1}-1)$$

$$= 8 \,.\, 5^{m-1}-2,$$

and

$$A_{2m+1} = 2+\sum_{n=1}^{m} 4\,.\,5^{n-1}+\sum_{n=1}^{m} 12\,.\,5^{n-1}$$

$$= 2+16\sum_{n=1}^{m} 5^{n-1}$$

$$= 2+4(5^m-1)$$

$$= 4\,.\,5^m-2.$$

Therefore, there are $8\,.\,5^{(k-2)/2}-2$ invertible numbers less than 10^k if k is even and $4\,.\,5^{(k-1)/2}-2$ if k is odd. In particular, there will have been 198 years like 1961 by A.D. one million, 22 of which have already passed.

REFERENCES

1.1.	*Some Notes on Elementary Arithmetic*	
	C. D. Langford	C170, Vol. 52, No. 381, p. 271
		1622, Vol. 26, No. 271, p. 184
	D. A. Lee	3040, Vol. 46, No. 358, p. 316
	N. M. Gibbins	1695, Vol. 27, No. 277, p. 216
1.2.	*Egyptian Mathematics*	
	R. S. Williamson	1786, Vol. 28, No. 282, p. 212
	D. G. Wells	C212, Vol. 54, No. 388, p. 144
1.3.	*Approximations to Square Roots*	
	H. Lindgren	2244, Vol. 35, No. 314, p. 252
	R. H. Macmillan	2909, Vol. 44, No. 349, p. 209
1.4.	*Using Pythagorean Number-Triples*	
	J. W. Holmes	C66, Vol. 45, No. 351, p. 49
	T. H. Beldon	C77, Vol. 45, No. 354, p. 334
1.5.	*Diagrams to illustrate the Geometric Series*	
	K. A. Deadman	C209, Vol. 54, No. 388, p. 140
	J. Gagan	2459, Vol. 38, No. 326, p. 281
1.6.	*1961*	
	J. F. Bowers	2974, Vol. 45, No. 354, p. 322

2 Probability and Combinatorics

In the middle of the 17th century, Blaise Pascal and Henri Fermat were corresponding over a dicing problem posed by a noted gamester, the Chevalier de Méré (the pair of dice that had been winning money for him had begun to lose it even faster!).†

Today, the child born on a gambling table has grown into a dignified and important member of the family of mathematics. It is essential in actuarial and insurance work and in all branches of mathematical statistics and some branches of modern physics.‡

Appropriately, the selection of problems in this section commences with some concerned with betting.

2.1 Beating the Bookies

The odds on the various runners in a race are a, b, c, ... to 1. Then, by suitably backing every runner, so long as one runner (no matter which) completes the course, a profit will be made if

$$\sum \frac{1}{a+1} < 1.$$

Two preliminary results are needed:

1. It is better to bet on a straight win rather than 'both ways'.

Proof. To guarantee a win, every runner must be backed. Now backing 'both ways' in effect increases the chances of backing a winner at the expense of lowering the overall odds.

But, if one backs every horse, one's chance of backing a winner can not be further increased.

Thus backing 'both ways' would merely reduce the odds with no compensating advantages.

Therefore we need only consider bets for a straight win.

2. In order to get back a sum £S (this includes returned stake money as well as profit) on a runner winning with odds a to 1 it is necessary to stake £$S/(a+1)$.

Proof. A stake of £$S/(a+1)$ at a to 1 gives a profit of £$aS/(a+1)$ together with original stake £$S/(a+1)$. Total £S.

† Huff, D., *How to take a chance*, Pelican, 1965, p. 52.

‡ Hooper, A., *Makers of Mathematics*, Faber, 1961, p. 227.

From these two results it follows that to guarantee a return of £S, whatever the winner, the total stake must be $\sum S/(a+1)$, the summation extending over all the odds. Thus there is certain unlimited profit if $\sum S/(a+1) < S$, i.e. if

$$\sum \frac{1}{a+1} < 1.$$

For one particular race, $\sum 1/(a+1) = 1{\cdot}0790$, a very close shave for the bookies!

2.2 Problem arising from the Crap Game

The game of craps is one of the most favoured of those involving dice. Further details of the odds may be found in Huff.†

The rules of the Crap Game are as follows. 'It is played with two ordinary dice. The man in possession of them wins immediately if the total of the spots on his first throw is 7 or 11 and loses immediately if this total is 2, 3 or 12; in either case he continues to throw the dice. If the total is any one of the remaining six possible values he continues to roll the dice until he has either duplicated his own first throw, in which case he wins, or has thrown a total of 7, in which case he loses and is then required to surrender the dice to the other player.'

It is reasonable to ask the question: how many times does a player expect to throw before handing the dice to the other player?

At any throw the probabilities of throwing the possible values are given by:

2	3	4	5	6	7	8	9	10	11	12
$\frac{1}{36}$	$\frac{2}{36}$	$\frac{3}{36}$	$\frac{4}{36}$	$\frac{5}{36}$	$\frac{6}{36}$	$\frac{5}{36}$	$\frac{4}{36}$	$\frac{3}{36}$	$\frac{2}{36}$	$\frac{1}{36}$

Suppose that the initially expected number of throws is u.

If the first throw is 4 or 10 let the expected number of subsequent throws be denoted by v_1.

† Huff, D., *How to take a chance*, Pelican, 1965, p. 58.

Similarly if the first throw is 5 or 9 let the expected number of subsequent throws be v_2 and if the first throw is 6 or 8 let the expected number of subsequent throws be v_3.

If the first throw is 2, 3, 7, 11 or 12 the game starts again at the second throw. In this case the expected number of subsequent throws is the same as the initially expected number and is therefore given by u.

To find the average expected number of throws subsequent to the first throw we must multiply the probability of the various possible first throws by the above expected numbers of subsequent throws and then add the products.

This gives for the average number of throws after the first

$$\tfrac{1}{6}v_1 + \tfrac{2}{9}v_2 + \tfrac{5}{18}v_3 + \tfrac{1}{3}u.$$

But we must have that the initially expected number of throws $= 1 +$ average number of throws after the first

i.e. $$u = 1 + \tfrac{1}{6}v_1 + \tfrac{2}{9}v_2 + \tfrac{5}{18}v_3 + \tfrac{1}{3}u.$$

In the same way we can consider the expected number of throws after the second throw. Suppose, first, that the first throw had been a 4. Then at the second throw there are three different possibilities:

> a throw of 4, with probability $\tfrac{1}{12}$, leads to a new start;
> a throw of 7, with probability $\tfrac{1}{6}$, ends the sequence of throws;
> all other throws, with probability $\tfrac{3}{4}$, leave the subsequently expected number of throws the same as before the second throw.

Again noting that the expected number of throws after the first throw is one greater than the average number after the first throw we have:

$$v_1 = 1 + \tfrac{1}{12}u + \tfrac{1}{6}\,.\,0 + \tfrac{3}{4}v_1.$$

Similarly $$v_2 = 1 + \tfrac{1}{9}u + \tfrac{1}{6}\,.\,0 + \tfrac{13}{18}v_2,$$

and $$v_3 = 1 + \tfrac{5}{36}u + \tfrac{1}{6}\,.\,0 + \tfrac{25}{36}v_3.$$

Solving the four simultaneous equations we find that $u =$ initially expected number of throws $= 8\tfrac{103}{196}$.

Games which depend more on skill than on chance but where, nevertheless, the 'luck of the draw' is influential, are represented by the next two examples on squash and the F.A. cup.

2.3 Squash Chances

In the game of squash a player wins a point if he serves and wins a rally, but not if he wins a rally in which he did not serve; the winner of a rally serves for the next. A game is won by the first player to obtain nine points (ignoring the slight change which may be made at 8-all). The question arises of the advantage given to the player who takes the first service, supposing the players to be of equal ability.

We take the phrase 'of equal ability' to mean that in any rally A's chances of winning are $\frac{1}{2}$ and B's are $\frac{1}{2}$. When A needs m points and B needs n points to win the game, call A's chances of winning the game $a_{m,n}$ when A is serving and $b_{m,n}$ when B is serving. We have immediately

$$\left.\begin{aligned} a_{m,n} &= \tfrac{1}{2}(a_{m-1,n}+b_{m,n}) \\ b_{m,n} &= \tfrac{1}{2}(a_{m,n}+b_{m,n-1}) \end{aligned}\right\}, \tag{1}$$

with the initial conditions $a_{0,n} = 1$, $b_{m,0} = 0\,(m, n \geqq 1)$. The symbols $a_{m,0}$, $b_{0,n}$ are undefined.

Using the generating functions

$$A \equiv A(u, v) \equiv \sum_{m=1}^{\infty} \sum_{n=1}^{\infty} a_{m,n} u^m v^n,$$

$$B \equiv B(u, v) \equiv \sum_{m=1}^{\infty} \sum_{n=1}^{\infty} b_{m,n} u^m v^n,$$

and multiplying each of the equations (1) by $u^m v^n$ and summing, we have by virtue of the initial conditions

$$(1-\tfrac{1}{2}u)A = \tfrac{1}{2}uv/(1-v)+\tfrac{1}{2}B,$$
$$(1-\tfrac{1}{2}v)B = \tfrac{1}{2}A,$$

so that on reduction

$$A = \frac{\tfrac{1}{2}uv(1-\tfrac{1}{2}v)}{(1-v)\{(1-\tfrac{1}{2}u)(1-\tfrac{1}{2}v)-\tfrac{1}{4}\}},$$

$$B = \frac{\tfrac{1}{4}uv}{(1-v)\{(1-\tfrac{1}{2}u)(1-\tfrac{1}{2}v)-\tfrac{1}{4}\}}.$$

Hence

$$A-B = \frac{\frac{1}{4}uv}{(1-\frac{1}{2}u)(1-\frac{1}{2}v)-\frac{1}{4}}.$$

Thus

$$A-B = \tfrac{1}{4}uv \sum_{r=0}^{\infty} (1-\tfrac{1}{2}u)^{-(r+1)}(1-\tfrac{1}{2}v)^{-(r+1)}(\tfrac{1}{4})^{r},$$

in which the coefficient of $u^m v^n$ $(m, n \neq 0)$ is

$$a_{m,n}-b_{m,n} = (\tfrac{1}{2})^{m+n} \sum_{r=0}^{\infty} \left\{ \frac{(m)_r(n)_r}{r!r!} (\tfrac{1}{4})^r \right\},$$

so that

$$a_{m,n}-b_{m,n} = (\tfrac{1}{2})^{m+n} F(m, n\,;\,1\,;\,\tfrac{1}{4}). \tag{2}$$

Since the players are of equal ability, we also have the relation

$$a_{m,n}+b_{n,m} = 1 \tag{3}$$

and it follows from (2) and (3) that

$$a_{m,m} = \tfrac{1}{2}+(\tfrac{1}{2})^{2m+1} F(m, m\,;\,1\,;\,\tfrac{1}{4}).$$

Replacing $F(m, m\,;\,1\,;\,\frac{1}{4})$ by $(\frac{4}{3})^{2m-1} F(1-m, 1-m\,;\,1\,;\,\frac{1}{4})$, we have in the case $m = 9$

$$a_{9,9} = \tfrac{1}{2}+\tfrac{1}{4}(\tfrac{2}{3})^{17} F(-8, -8\,;\,1\,;\,\tfrac{1}{4})$$

which may be calculated, since the series terminates. We obtain

$$a_{9,9} = \frac{7\,674\,706}{14\,348\,907}.$$

2.4 Football Probabilities

During the broadcast of the draw for the 16 matches in the fourth round of the F.A. Cup (1957), the commentator, Mr. Raymond Glendenning, said he thought that, since 6 of the teams were from London,† there was a good chance of a match between 2 London teams. It is interesting to investigate what the chance is. We consider first the general case, where there are $2n$ teams, of which m are from London.

† Followers of the game may recall that one of the matches was Tottenham Hotspur v. Chelsea.

Throughout we take account of the order of the n pairs of opposing teams, but we ignore the order in each pair, that is, we ignore the distinction between home and away teams. The total number of ways in which the $2n$ teams can be paired off is $(2n)!/2^n$.

We now calculate the number of ways in which the draw can be made so that no 2 London teams play each other. The number of ways of placing the m London teams in the n pairs, with no 2 in one pair, is $n!/(n-m)!$, the number of ways of choosing m of the $2n-m$ non-London teams to play against them is $(2n-m)!/(2n-2m)!$, and the number of ways of pairing off the other $2n-2m$ non-London teams is $(2n-2m)!/2^{n-m}$. Hence the number of ways is $n!(2n-m)!/(n-m)!2^{n-m}$. The probability that no 2 London teams play each other is, therefore,

$$\frac{n!(2n-m)!2^m}{(n-m)!(2n)!},$$

that is,

$$\frac{(2n-m)(2n-m-2)\ldots(2n-2m+2)}{(2n-1)(2n-3)\ldots(2n-m+1)}$$

if m is even, and

$$\frac{(2n-m-1)(2n-m-3)\ldots(2n-2m+2)}{(2n-1)(2n-3)\ldots(2n-m+2)}$$

if m is odd.

In the given case, where $m = 6$ and $n = 16$, this is 0·5656 (to 4 places). Hence the probability of at least one match between 2 London teams is 0·4344. The probabilities of 1, 2 or 3 such matches may be calculated separately. The results are 0·3856, 0·0482, and 0·0006.

Problems about selections, arrangements and runs are interesting in themselves and for the contribution which they make to the development of probability theory. One is reminded by the subject of 2.4 that much present-day gambling is connected with 'doing the Pools' and that permutations (combinations) to help you, and everyone else, win are made freely available by the Pools firms!

2.5. Runs of Successes and Failures

A psychologist often wishes to know whether a long run of correct responses by his rat could merely be due to chance. In investigating this problem, the result below was found. This is submitted as a piece of pure mathematics; it is not well suited to the practical problem.

Suppose a penny is tossed $r+x$ times. How many runs of exactly r heads are to be expected?

If the run of exactly r heads occurs in the first r tossings, the $(r+1)$th tossing must give a tail. There are 2^{x-1} different ways in which the remaining tosses can result. The same number is obtained if the last r tosses yield the run. For intermediate positions, the run of heads must both be preceded and followed by a tail. There are 2^{x-2} ways for the remaining tosses.

Thus, if x is so small that it is impossible for two runs of r heads to occur, out of the 2^{r+x} possible outcomes, there will be $2 \,.\, 2^{x-1} + (x-1) \,.\, 2^{x-2}$ cases in which a run of exactly r heads occurs. Thus we have $(x+3) \,.\, 2^{x-2}$ cases, provided x is at least 1. If $x = 0$, there is of course just one way in which the run can occur.

But the above argument is still useful for larger x. If we imagine the 2^{r+x} possible cases written out, the argument tells us that there will be 2^{x-1} cases in which tosses 1 to r constitute a run, 2^{x-2} in which 2 to $(r+1)$ constitute a run, and so on. If a case contains two runs, it will appear twice in the above list; if three, three times. But this is just what should happen if we are attempting to determine the total number of runs in the 2^{r+x} possible cases.

Thus, for all x greater than 0, the total number of runs of precisely r heads in the 2^{r+x} possible cases for $(r+x)$ tossings is $(x+3) \,.\, 2^{x-2}$. Beginning from $x = 0$, the early numbers of the sequence thus defined are 1, 2, 5, 12, 28, 64.

2.6 Permutation Problems

I A lesson on elementary permutations produced the following problem from a pupil : 'If my C.C.F. squad comprises n pupils, for how many weeks can they parade in such a way that no pupil has to stand next to any other pupil more than once, the whole squad parading each time in a single rank ?'

If we call each instance of two cadets being adjacent a 'pairing', the total number of different pairings amongst n pupils is $n(n-1)/2$.

Since $n-1$ pairings occur at each parade, they certainly cannot parade for more than $\frac{1}{2}n$ weeks.

If n is even, the upper bound of $\frac{1}{2}n$ is always attainable. In the following demonstration of this, a set of lines is constructed satisfying the conditions. The solution is shown in the accompanying diagram for $n = 8$, but it is clearly equally applicable to any even number of pupils.

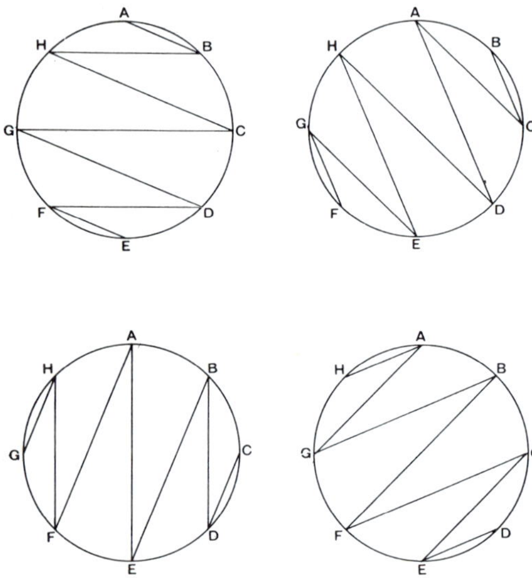

The order in which the points are encountered along the broken line gives the order of pupils in the rank. Thus this particular solution gives the order of pupils on parade for 4 weeks as follows: *ABHCGDFE*, *BCADHEGF*, *CDBEAFHG*, *DECFBGAH*. Each cadet will parade at the end of the line just once.

If n is odd, there cannot be more than $\frac{1}{2}(n-1)$ parades, since this is the greatest integer not greater than $\frac{1}{2}n$. The arrangements of the pupils in this case may be constructed by setting down the arrangements for $n-1$ cadets and then placing the last cadet at the end of the line in each rank; he will then have a different neighbour each time.

We deduce that the solution of the problem in every case is $[\frac{1}{2}n]$.

II *Method of attack for problems about permutations*

If 4 pupils A, B, C, D are to be arranged in a row, i.e., in positions 1, 2, 3, 4 which may be imagined as vacant cells ready to receive them, there are two well-known methods of showing that the number of arrangements is 4!.

1. *Fill the cells successively*: There is a choice of 4 pupils for cell 1, then 3 for cell 2, etc.
2. *Allot the pupils successively*: There is a choice of 4 cells for A, then 3 for B, etc.

However, a third method of proof exists but it is not so widely recognised.

3. *Assign relative positions to the pupils successively*: You must clear your mind of any idea of cells and imagine A standing on parade; B joins him with a choice of 2 positions (left to right), for each of which C then has a choice of 3 positions (left, right, or between A and B), and finally D has a choice of 4 'gaps'. This word 'gap' is used here to mean either a genuine gap between 2 pupils or the position beyond an end of an existing rank.

This third method gives a *natural* proof of the result $8 \times 9!$ for the number of ways of arranging 10 pupils in a rank if a specified 2 of them must not be adjacent (we shall say they are a taboo pair). Proofs by other methods are longer and less satisfying.

The 'other' 8 and one member of the taboo pair may be arranged in 9! ways; then obviously 2 out of the 10 potential 'gaps' are not available to the other taboo member, hence the total number of arrangements is $8 \times 9!$

Similar problems with ONE larger taboo group yield easily to this method. For example, if 4 out of 10 form a taboo group then there are $7!6 . 5 . 4$ arrangements; proofs of this by other methods are very long. But unfortunately problems with more than one taboo group do NOT yield easily to this method. For one delirious moment the writer thought that the number of ways of arranging 10 pupils including 2 taboo pairs (each of 2) was $8!7 . 8$; this was based on an argument '8! for the 6 normal and 1 from each pair, then 7 (out of the 9 gaps) for the next, and 8 (out of 10 gaps) for the last'. But this is fallacious. No doubt readers will see why.

III *A Rearrangement Analogy*

In the *Answers and Hints for Books 1 and 2* of SMP *Advanced Mathematics*, page 165, there is a note concerning the symmetries of the regular tetrahedron *ABCD*. It is pointed out that two interpretations can be given for an expression such as

$$\begin{pmatrix} 1 & 2 & 3 & 4 \\ 2 & 3 & 1 & 4 \end{pmatrix}.$$

The first definition signifies that the vertex currently in position 1 moves to position 2, the vertex in position 2 to position 3, etc. This is the definition used in the SMP text. Following the article by Mr. D. B. Hunter,† we call this a *rearrangement*, and denote it by

$$p_1' = \begin{Bmatrix} 1 & 2 & 3 & 4 \\ 2 & 3 & 1 & 4 \end{Bmatrix}.$$

The second definition signifies that vertex *A* moves to the position occupied by vertex *B*, vertex *B* to the position occupied by vertex *C*, etc. This is called a *permutation*, and will be denoted by

$$p_1 = \begin{pmatrix} 1 & 2 & 3 & 4 \\ 2 & 3 & 1 & 4 \end{pmatrix}.$$

The SMP answer book suggests that the interesting property‡

$$p_1'p'_2 = p_2p_1$$

is difficult to explain. This note suggests an analogy which may be helpful.

Suppose that we have four balls, four boxes and four holes, each numbered 1 to 4. We consider two ways in which the balls can be placed inside the boxes which in turn are inside the holes. The arrangement considered as an example is illustrated in Fig. 1.

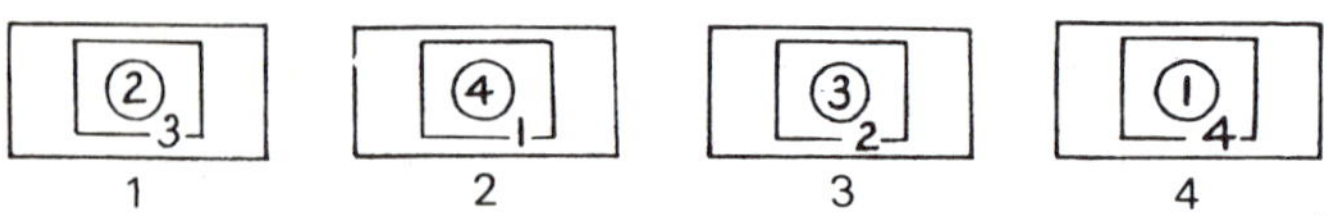

FIG. 1

† Hunter, D. B., Permutations and Rearrangements, *Mathematical Gazette*, Vol. 50, No. 373, p. 290.

‡ We use the convention that the transformation performed first is written last.

First, we may place the balls 1, 2, 3, 4 in the boxes 4, 3, 2, 1 and then place the boxes 1, 2, 3, 4 in the holes 2, 3, 1, 4. This corresponds to combination of the rearrangements

$$p_2' = \begin{Bmatrix} 1 & 2 & 3 & 4 \\ 4 & 3 & 2 & 1 \end{Bmatrix} \quad \text{and} \quad p_1' = \begin{Bmatrix} 1 & 2 & 3 & 4 \\ 2 & 3 & 1 & 4 \end{Bmatrix};$$

in that order; that is, to

$$p_1'p_2' = \begin{Bmatrix} 1 & 2 & 3 & 4 \\ 4 & 1 & 3 & 2 \end{Bmatrix}.$$

Fig. 2 illustrates the interpretation, first in terms of transformations of the tetrahedron and then in terms of the mapping of balls to boxes and boxes to holes.

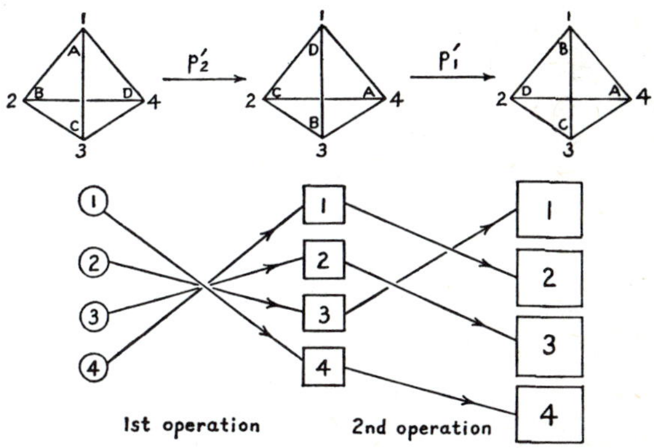

FIG. 2

Alternatively, we may begin by placing the boxes 1, 2, 3, 4 in the holes 2, 3, 1, 4 and then place the balls 1, 2, 3, 4 in the boxes 4, 3, 2, 1 (which have now, of course, moved into new positions). This corresponds to combination of the permutations

$$p_1 = \begin{pmatrix} 1 & 2 & 3 & 4 \\ 2 & 3 & 1 & 4 \end{pmatrix} \quad \text{and} \quad p_2 = \begin{pmatrix} 1 & 2 & 3 & 4 \\ 4 & 3 & 2 & 1 \end{pmatrix};$$

that is, to

$$p_2 p_1 = \begin{pmatrix} 1 & 2 & 3 & 4 \\ 4 & 1 & 3 & 2 \end{pmatrix}.$$

The appropriate interpretation is illustrated in Fig. 3.

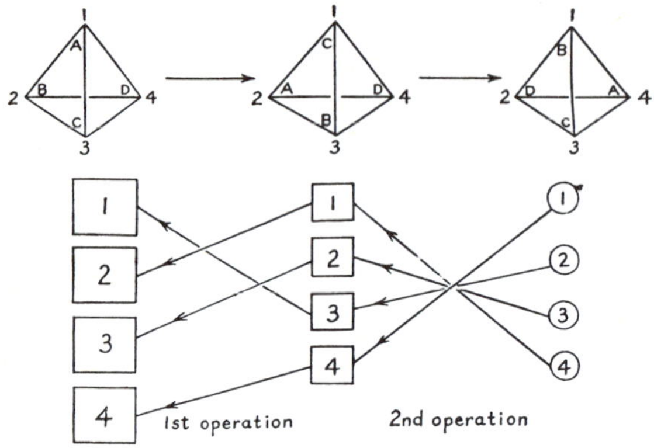

FIG. 3

Since it is immaterial whether the boxes are placed in the holes before or after the balls are placed in the boxes, it follows that

$$p_1'p_2' = p_2p_1.$$

2.7 Problems of Paths

I A familiar topological puzzle asks for the number of paths between two marked points connected by a maze of intersecting lines. More precisely, a finite number of points are joined, each to each, by line segments, and the problem is to find the number of simple contours, formed of line segments, connecting one chosen point with another. We shall confine ourselves here to the simplest case in which each line segment may be supposed to carry an arrow indicating the unique direction in which this segment may be traversed; in practice this restriction is effected by prohibiting any move which has a positive component *towards* the starting point.

An obvious solution is obtained by assigning, *progressively*, to each point of the configuration the number of the totality of paths leading to it from the starting point, in the following manner:

The starting point is numbered 1; any other point bears the number which is the sum of the numbers at the far ends of all the line segments *leading* to this point.

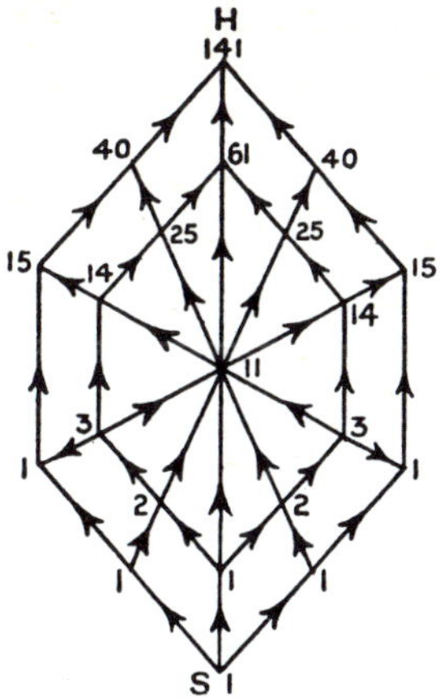

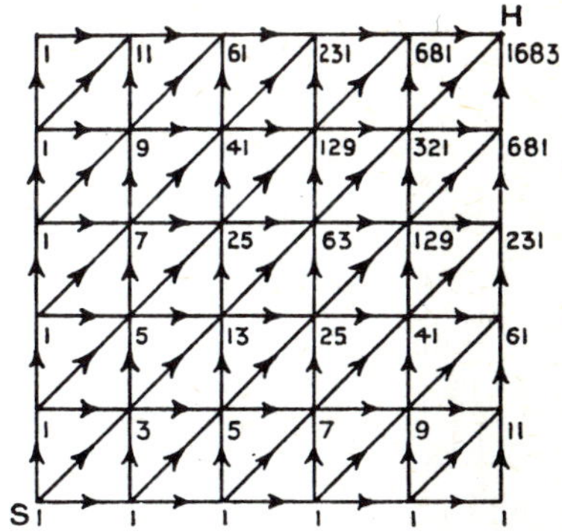

The method is illustrated in the two diagrams; in each S denotes the start, and H is home.

II Towns A and B are at opposite vertices of a square S. A network of roads divides S into n^2 equal sub-squares, of unit side, each side of S being a road. To find the number of paths, from A to B, of length $2n$. Let l be one side of S which goes through A; $n+1$ roads lead off l; suppose these roads are numbered from 1 to $n+1$. Let x_r be the distance travelled along the rth road in a journey from A to B, $x_r = 0$ denoting that the rth road is not taken. Then the number of paths of length $2n$ is the number of solutions in non-negative integers of the equation

$$x_1+x_2+\ldots+x_{n+1} = n.$$

This is the same as the number of ways of choosing n out of $2n$ objects (for if we choose n out of $2n$ we leave n divided into $n+1$ groups, some of which may be empty). Thus the number of paths is ${}^{2n}C_n$.

III 'The boundary of Squaretown is a square whose sides, running north-south or east-west, are of length $na+(n-1)d$. The town is divided into n^2 blocks of side a by $(n-1)$ streets from north to south and $(n-1)$ streets from east to west, each street being of width d. There is also a street—of unspecified width—encircling the whole town. What is the shortest distance from the extreme north-east corner to the extreme south-west corner of the town, and how many possible routes are of this minimum length?' Two possible routes are shown.

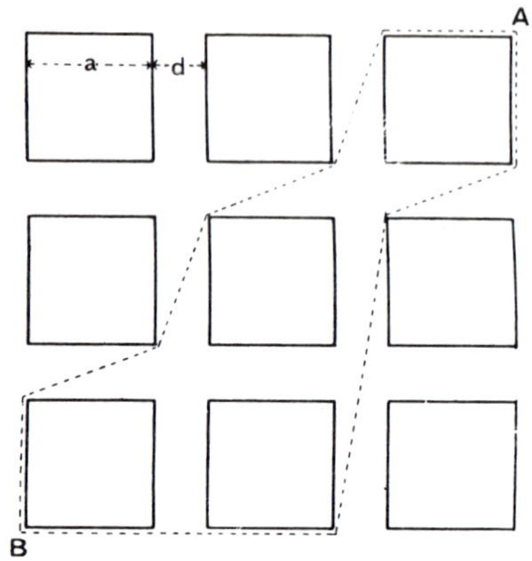

IV *Cyclically Symmetric Knight-tours*

How many cyclically symmetric knight-tours on a board ten by ten? It appears that the only method of arriving at the exact answer is akin to that put forward by an American big business man when asked by a somewhat over-subtle English efficiency expert how he would ascertain precisely what percentage of his labour force was productively employed—'Count their durned heads'.

No human computer is likely to have the time or the patience to make the count, but an estimate is now given, after classifying the tours into types and counting a representative selection. Here are four of the many beautiful results which emerge from this investigation.

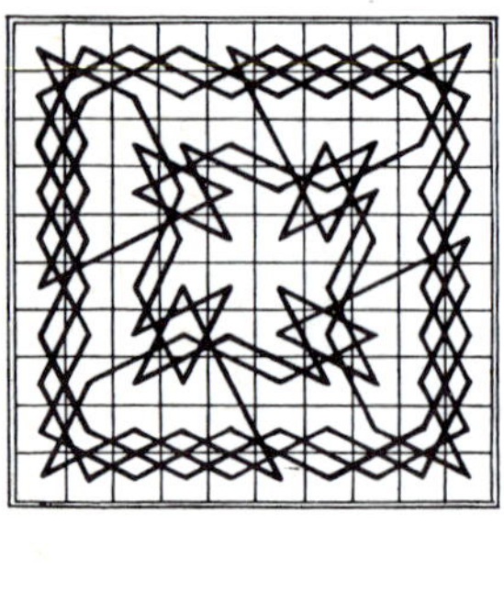

The total seems to be in the neighbourhood of 200 000—a surprisingly large number perhaps when one remembers that the number on the only smaller board susceptible to such tours—the six by six—is precisely 5. The next bigger board—fourteen by fourteen—would presumably give hundreds of millions.

2.8 Transport Problems

'I am at a corner where I have the choice of two independent bus systems, buses on each running at ten-minute intervals. How long may I expect to wait for a bus?'

We assume that the waiting times x, y for the buses have the rectangular distribution over the interval $(0, 10)$, so that their probability density functions are $f(x) = g(y) = \frac{1}{10}$.

The chance that *neither* comes in the first t minutes is $(1-t/10)^2$ and so the chance of getting a bus in t minutes is $1-(1-t/10)^2$. Hence if $h(u)$ is the density function of $u \equiv \min(x, y)$, we have

$$\int_0^t h(u)\,du = 1-\left(1-\frac{t}{10}\right)^2,$$

whence

$$h(u) = 2\,.\,\frac{1}{10}\,.\left(1-\frac{u}{10}\right) = \frac{10-u}{50}$$

so that $E(h) = \displaystyle\int_0^{10} uh(u)\,du = 3\tfrac{1}{3}$ minutes.

In the case of n independent bus systems we would have

$$\int_0^t h(u)\,du = 1-\left(1-\frac{t}{10}\right)^n$$

and therefore

$$h(u) = \frac{n}{10}\left(1-\frac{u}{10}\right)^{n-1}$$

so that

$$E(u) = \frac{n}{10^n}\int_0^{10} u(10-u)^{n-1}\,du = 10nB(2,n) = \frac{10}{n+1}\text{ minutes.}$$

The following extensions are suggested.

Problem 1. I arrive at a bus stop where I have a choice of n independent bus systems, on each of which buses run at intervals T. How long may I expect to wait for a bus?

Taking the instant of my arrival as the origin of time t, the probability that the next bus of a specified system will arrive later than time t is $(1-t/T)$. Thus the probability that the waiting time exceeds t is

$$Q_n(t) = (1-t/T)^n, \quad 0 \leqq t \leqq T.$$

Hence the probability that the waiting time lies between t and $t+dt$ is

$$P_n(t)\,dt = -Q_n'(t)\,dt = nT^{-1}(1-t/T)^{n-1}\,dt,$$

and the expectation of the waiting time is

$$\bar{t} = \int_0^T tP_n(t)\,dt = \int_0^T Q_n(t)\,dt = T/(n+1).$$

Problem 2. This is the same as Problem 1, but with the additional information that no bus has passed between time $-u$ and the present.

With an obvious modification in the above notation,

$$Q_n(t,u) = \{1-t/(T-u)\}^n, \quad 0 \leqq t \leqq T-u,$$
$$P_n(t,u)\,dt = -(d/dt)Q_n(t,u)\,dt,$$
$$\bar{t} = (T-u)/(n+1).$$

Problem 3. This is the same as Problem 2, but with the additional information that a bus passed at the instant $-u$.

The solution is the same as in Problem 2, but with n replaced by $n-1$, since a bus from each of the $n-1$ systems will arrive before time $T-u$. If $n = 1$ there is only one possible waiting time $T-u$.

Matching problems occupy a fair amount of space in various works on probability and provide useful illustrations of combinatorial theory at various levels.

Perhaps the simplest is the proof of the result that if we have more than 23 people at a party then the probability that at least 2 have the same birthday is greater than $\frac{1}{2}$ (but note that the probability of a person's brithday falling on any one day of the year is *not* equal to that for any other day). Other examples follow.

2.9 How Unlikely?

Two shuffled packs of playing cards are placed side by side, face downwards. Two cards, one from each pack, are drawn repeatedly, no card being replaced. To find the chance that at the same draw the two cards are of the same suit and same denomination.

The problem is equivalent to finding the chance that if 52 numbered objects are placed at random in 52 numbered holes, at least one object is in its right hole, that is, hole and object have the same number. The solution is very simple, but the result rather unexpected.

Consider the case of n holes and n objects. Let p_n be the number of arrangements in which none of the n objects is in its right hole. Then the number of arrangements in which r and only r objects out of n are in their right holes is ${}^nC_r \,.\, p_{n-r}$.

Hence

$$\sum_{r=1}^{n} {}^nC_r \,.\, p_{n-r} = n! - p_n$$

or

$$\sum_{r=0}^{n} {}^nC_r \,.\, p_{n-r} = n!.$$

Denote the left-hand side by a_n, and consider

$$\sum_{r=0}^{n} (-1)^r \, {}^nC_r \,.\, a_{n-r},$$

where a_0 stands for 1.

Now $$\begin{aligned}{}^{n}C_r - {}^{n}C_1 \,.\, {}^{n-1}C_{r-1} + {}^{n}C_2 \,.\, {}^{n-2}C_{r-2} - \ldots + (-1)^r\, {}^{n}C^r \\ = {}^{n}C_r\{1 - {}^{r}C_1 + {}^{r}C_2 \ldots + (-1)^r\} \\ = {}^{n}C_r(1-1)^r = 0.\end{aligned}$$

Thus

$$\sum_{r=0}^{n} (-1)^r\, {}^{n}C_r \,.\, a_{n-r} = p_n.$$

But $a_n = n!$ and $\sum_{r=0}^{n} (-1)^r\, {}^{n}C_r \,.\, (n-r)! = \sum_{r=0}^{n} (-1)^r n!/r!$.

Thus

$$p_n = n!\left\{\frac{1}{2!} - \frac{1}{3!} + \frac{1}{4!} \ldots + \frac{(-1)^n}{n!}\right\},$$

and so the probability required is $1 - p_n/n! = 1 - e^{-1}$ approximately, with an error of less than $1/(n+1)!$.

When $n = 52$, the chance is therefore $1 - e^{-1}$ with an error of less than $1/53!$; the chance is roughly 17/27.

2.10 The Hat Check Problem

'Four men dined at the Ambrosia. They parked their hats in the cloakroom. Each of the four diners came away wearing a hat belonging to one of the others. In how many ways could this have happened?'

It is not difficult to see that the answer is 9. But the mathematician will naturally wish to discover the general rule. What will be the answer for n men? Call the number of ways u_n, and the diners A, B, C, ... If A interchanges hats with B, the remaining $n-2$ can each wear a wrong hat in u_{n-2} ways. The result is the same if A interchanges with C. Therefore if A interchanges with *one* of the others the number of ways is covered by $(n-1)u_{n-2}$. If A wears B's hat but B does not wear A's, the number of ways is u_{n-1} (for, as far as B is concerned, A's hat is barred to him, this possibility having already been allowed for). So we get $(n-1)u_{n-1}$ more ways. This exhausts the possibilities. So we have the equation:

$$u_n = (n-1)(u_{n-1} + u_{n-2}).$$

This is a well-known type of difference equation, though it is not easy to come across illustrations of its use.

It can be written:

$$u_n - nu_{n-1} = -(u_{n-1} - \overline{n-1}\,u_{n-2}).$$

Stepping down we have:

$$u_{n-1} - \overline{n-1}u_{n-2} = -(u_{n-2} - \overline{n-2}u_{n-3}),$$

$$\dots\dots\dots\dots\dots\dots\dots\dots$$

down to $u_4 - 4u_3 = -(u_3 - 3u_2)$.

From the conditions of the question, $u_2 = 1$, $u_3 = 2$. Hence $u_3 - 3u_2 = -1$. So we have $u_n - nu_{n-1} = (-1)^n$.

This may be written:

$$\frac{u_n}{n!} - \frac{u_{n-1}}{(n-1)!} = \frac{(-1)^n}{n!}.$$

So

$$\frac{u_{n-1}}{(n-1)!} - \frac{u_{n-2}}{(n-2)!} = \frac{(-1)^{n-1}}{(n-1)!},$$

$$\dots\dots\dots\dots\dots\dots\dots\dots$$

$$\frac{u_3}{3!} - \frac{u_2}{2!} = \frac{(-1)^3}{3!},$$

$$\frac{u_2}{2!} = \frac{1}{2!}.$$

Whence

$$u_n = n!\left\{\frac{1}{2!} - \frac{1}{3!} + \frac{1}{4!} - \frac{1}{5!} \dots (-)^n \frac{1}{n!}\right\}.\dagger$$

We can calculate values of u_n from this formula. But for small values of n it is simpler to use the equation itself, i.e.

$$u_n = (n-1)(u_{n-1} + u_{n-2}).$$

Thus:

$u_2 =$	1	
$u_3 =$	2	$u_2 + u_3 = 3$
$u_4 =$	9	$u_3 + u_4 = 11$
$u_5 =$	44	$u_4 + u_5 = 53$
$u_6 =$	265	$u_5 + u_6 = 309$
$u_7 =$	1854	$u_6 + u_7 = 2119$
$u_8 =$	14 833	$u_7 + u_8 = 16\ 687$
$u_9 =$	133 496	$u_8 + u_9 = 148\ 329$
$u_{10} =$	1 334 961	

† The solution on these lines of a similar equation will be found in: Durell, C. V. and Robson, A., *Advanced Algebra*, 2, Bell, 1937, p. 232.

The next problems, one classical and one somewhat idealised example from the world of sport are included here as representatives of the class of problems which readily yield solutions when approached geometrically.

2.11 A Broken Stick

The following is a proof for the fact that if a rod is broken into three pieces, the probability that a triangle can be made from them is $\frac{1}{4}$.

The important feature is that no one of the three pieces exceeds the sum of the rest, which in itself implies that each length should be less than half the length of the rod.

Suppose that the length of the rod is a and the 3 pieces are of

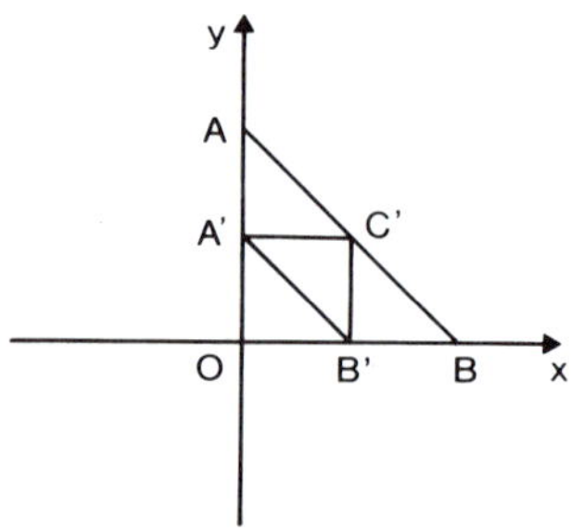

length x, y, z respectively. Now let x and y represent the axes of coordinates in 2 dimensions. All the ways of breaking the rod are represented by points inside and on the triangle formed by the axes and the line $x+y=a$, that is triangle OAB. The points representing ways of breaking the stick which do satisfy the condition are represented by points inside the area represented by the lines $x=\frac{1}{2}a$, $y=\frac{1}{2}a$ and $x+y \geqq \frac{1}{2}a$, that is triangle $A'B'C'$. The area of triangle AOB is $\frac{1}{2}a^2$ and that of triangle $A'B'C'$ is $a^2/8$.

The probability of the condition being satisfied is

$$\frac{a^2}{8}\bigg/\frac{a^2}{2} = \tfrac{1}{4}.$$

This problem admits of extension to the situation where a stick is broken into n pieces and the probability that a polygon of n or fewer angles can be made out of them is required.

A method is given below for four pieces and extended to n pieces.

Suppose that the rod is of length a and the lengths of the four separate pieces in a particular break are x, y, z, w. Then if this break is to satisfy the condition, it is necessary and sufficient that the length of each of the pieces should be not greater than half the length of the rod. If x, y, z are taken as axes of coordinates in three dimensions, all ways of breaking the rod are represented by points inside and on the tetrahedron formed by the coordinate planes and the plane $x+y+z = a$.

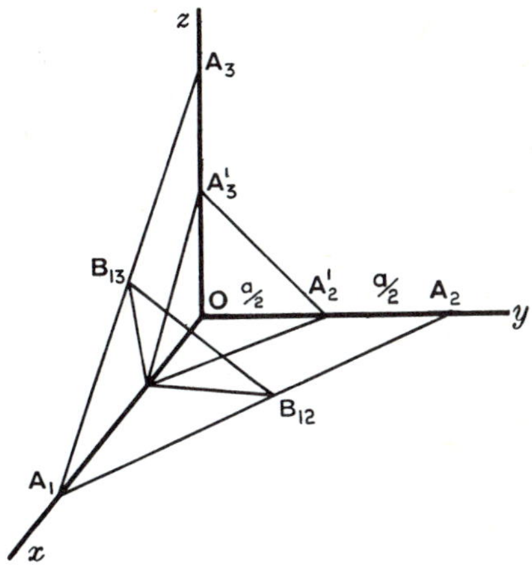

The points representing ways of breaking which satisfy the condition are represented by points inside the volume bounded by the planes $x = \frac{1}{2}a$, $y = \frac{1}{2}a$, $z = \frac{1}{2}a$, $x+y+z = \frac{1}{2}a$, $x+y+z = a$. If A_1 is the point on Ox such that $OA_1 = a$, A_1' the point on Ox such that $OA_1' = \frac{1}{2}a$, A_2 the point on Oy such that $OA_2 = a$, and so on, B_{12} the mid-point of A_1A_2 and so on, then the points not satisfying the required condition lie within the tetrahedron $OA_1'A_2'A_3'$, or within the three equal tetrahedra, on each axis, such as $A_1'A_1B_{12}B_{13}$. The volume of the tetrahedron representing all ways of breaking the rod is $a^3/6$, and the volume occupied by the points not satisfying the condition is $4a^3/48$, hence the required value is

$$1-(4 \,.\, 6/48) = \tfrac{1}{2}.$$

In the case of breaking into n pieces, we have a figure in $n-1$ dimensions. The volume of the polytope including the points representing all ways of breaking the rod is $a^{n-1}/(n-1)!$. The unfavourable ways are contained in n polytopes, one at the origin and one on each axis, as in the three-dimensional case, the volume of each of these being $a^{n-1}/2^{n-1}(n-1)!$. Thus the required probability is

$$1-n/2^{n-1}.$$

For $n = 1$, $n = 2$, this gives the correct probability zero; for $n = 3$ and $n = 4$ it gives the values $\frac{1}{4}$ and $\frac{1}{2}$ respectively.

2.12 A Case of Local Probability

A golf ball, driven from the tee, comes to rest on a perfectly square and level green, with no banks; the hole being exactly in the centre of the green. Nothing being assumed as to the player's skill, what is the chance that he will find his ball nearer to the pin than to any edge of the green?

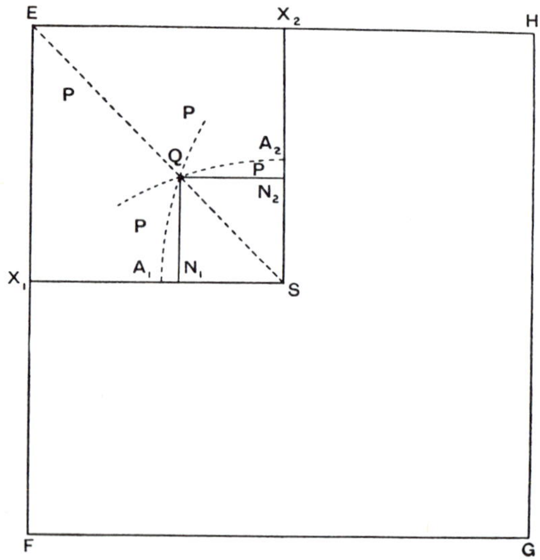

Let S be the centre of a square $EFGH$; let X_1, X_2 be the mid-points of the sides EF, HE; and let P be any point in the quarter square EX_1SX_2, which area suffices for the purposes of the problem. We have to find the chance that P is nearer to S than to either EX_1 or X_2E.

Bisect SX_1, SX_2 in A_1, A_2; and with vertices A_1, A_2, and common focus S, suppose two parabolas described towards E, intersecting on SE (as they must) in Q, say. Then if P lies within the space A_1QA_2S it is nearer to S than to either of the directrices EX_1, EX_2; but if it is outside the space A_1QA_2S, it is nearer to one or other directrix, or possibly to both, than to S. The chance of equidistance, when P is on either arc A_1Q, A_2Q, may be disregarded.

From Q let fall perpendiculars QN_1, QN_2 on SX_1, SX_2; and denote A_1S by 1; A_1N_1 by x; QN_1 by y.

Then $QN_1^2 = 4A_1N_1 \,.\, A_1S = 4x$.

Also $QN_1^2 = \frac{1}{2}SQ^2$; but $SQ = N_1X_1$; $\therefore\ QN_1^2 = \frac{1}{2}X_1N_1^2$
$= \frac{1}{2}(1+x)^2$; $\therefore\ x^2-6x+1 = 0$; from which $x = 3-2\sqrt{2}$.

Area $A_1QN_1 = \frac{2}{3}xy = \frac{4}{3}x\sqrt{x} = \frac{4}{3}(3-2\sqrt{2})^{3/2} = \frac{4}{3}(\sqrt{2}-1)^3$
$= \frac{4}{3}(5\sqrt{2}-7)$;
" A_2QN_2 is the same;

Area $QN_1SN_2 = QN_1^2 = 4x = 4(3-2\sqrt{2})$;
" $EX_1SX_2 = 4$.

Hence the required chance

$$= [(A_1QN_1)+(A_2QN_2)+(QN_1^2)]/EX_1SX_2 = \tfrac{2}{3}(5\sqrt{2}-7) + (3-2\sqrt{2}) = 0{\cdot}2189,$$

i.e. the chance is approximately 127/576, being greater than 7/32 but less than 2/9.

Finally a note which may well provoke a lot of discussion!

2.13 Randomness

The difficulty of interpreting the term 'at random' is well illustrated in the problem of finding the mean length of a large number of chords drawn at random in a circle of unit radius.

In this case 'at random' may be interpreted in a number of ways. Three examples are given below. Each of these might be intellectually satisfying if the others did not exist.

1. Consider all chords perpendicular to a given diameter. Then since the chords are drawn at random we may assume that they cut the diameter with uniform density and the mean chord length is:

$$2\int_0^1 \sqrt{1-x^2}\,dx = \tfrac{1}{2}\pi = 1{\cdot}57$$

2. Consider all chords drawn through a given point on the circumference. We may assume that they are uniformly distributed in angles to the diameter. This is equivalent to assuming that the other ends of the chords are uniformly distributed about the circumference. In this case, the mean chord length is:

$$\frac{4}{\pi}\int_0^{\frac{1}{2}\pi} \cos\theta\,d\theta = \frac{4}{\pi} = 1{\cdot}27$$

3. Apart from the centre, every point in a circle is the mid-point of a unique chord. There is, therefore, a one–one correspondence between the points and chords in a circle and selecting chords at random is equivalent to selecting chord centres at random. If we assume that random chord centres are uniformly distributed in the circle, the mean chord length is:

$$4\int_0^1 x\sqrt{1-x^2}\,dx = \tfrac{4}{3} = 1{\cdot}33$$

REFERENCES

2.1. *Beating the Bookies*
P. G. T. Lewis 2933, Vol. 44, No. 350, p. 285

2.2. *Problem Arising from the Crap Game*
J. C. Barton 2638, Vol. 40, No. 334, p. 276

2.3. *Squash Chances*
H. G. ApSimon 2697, Vol. 41, No. 336, p. 136

2.4. *Football Probabilities*
E. J. F. Primrose 2741, Vol. 42, No. 339, p. 37

2.5. *Runs of Successes and Failures*
W. W. Sawyer 2671, Vol. 41, No. 335, p. 54

2.6. *Permutation Problems*
D. A. Quadling 2527, Vol. 39, No. 328, p. 150
H. M. Finucan C235, Vol. 54, No. 390, p. 381
W. G. Rosser C263, Vol. 56, No. 396, p. 118

2.7. *Problems of Paths*
R. L. Goodstein 1913, Vol. 30, No. 290, p. 156
1364, Vol. 23, No. 254, p. 206
D. J. Finney 1423, Vol. 23, No. 257, p. 467
W. H. Cozens 2884, Vol. 44, No. 348, p. 117

2.8. *Transport Problems*
P. W. M. John 2723, Vol. 41, No. 338, p. 275
2814, Vol. 42, No. 342, p. 307

2.9. *How Unlikely?*
R. L. Goldstein 1363, Vol. 23, No. 254, p. 205

2.10. *The Hat Check Problem*
H. E. Piggott 2063, Vol. 33, No. 305, p. 214

2.11. *A Broken Stick*
G. A. Bull 2016, Vol. 32, No. 299, p. 87

2.12. *A Case of Local Probability*
G. Wotherspoon 955, Vol. 15, No. 207, p. 112

2.13. *Randomness*
E. E. Wright 2746, Vol. 42, No. 339, p. 43

3 Algebra

3.1 Solution of Quadratic Equations by Geometrical Constructions

A quadratic equation can be solved by a geometrical construction using ruler and set-square only.

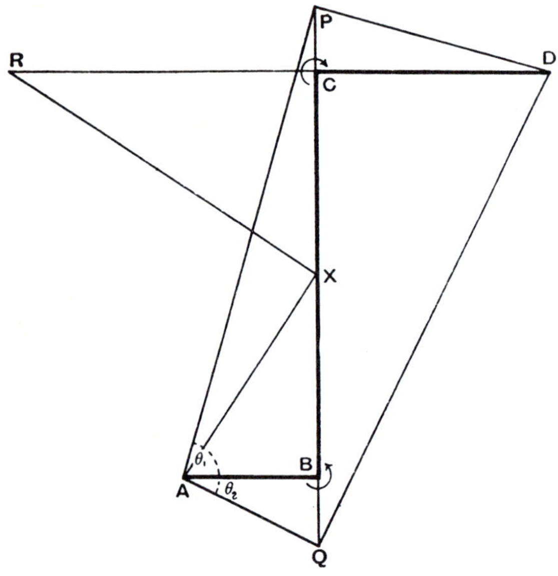

For example, let the equation be $2x^2-6x-3{\cdot}5=0$. The coefficients 2, 6, 3·5 are plotted successively at right angles, clockwise to each other, save when a change of sign occurs: $AB=2$ units, $BC=6$ units, $CD=3\frac{1}{2}$ units. Then P, Q are found on BC such that $\angle APD$, $\angle AQD$ are right angles. This can be done by set-square, or by drawing the circle on AD as diameter. The latter method will give greater accuracy.

Now $\quad BP = 6+CP = 2\tan\theta_1,$

$$CD = 3{\cdot}5 = PC\tan\theta_1 = (2\tan\theta_1-6)\tan\theta_1,$$

and hence $3{\cdot}5 = 2\tan^2\theta_1 - 6\tan\theta_1$.

Thus the roots are $\tan\theta_1$, $\tan\theta_2$; in this case 3·5, −0·5.

The minimum value of the expression is given when θ is the angle which XA makes with AB, where X is the mid-point of BC; in this case $\tan\theta = 1{\cdot}5$. The value of the expression is then given by RD, where XR is at right angles to AX.

A simpler construction was given by K. Wigand.†

To solve the equation $x^2+px+q=0$ with real coefficients p, q we draw the circle on the line joining the points $(0, 1)$, $(-p, q)$ as diameter; the x-coordinates of the points where this circle meets the x-axis are the roots of the given quadratic.

This can be proved quite easily by coordinate geometry by finding the equation of the circle.

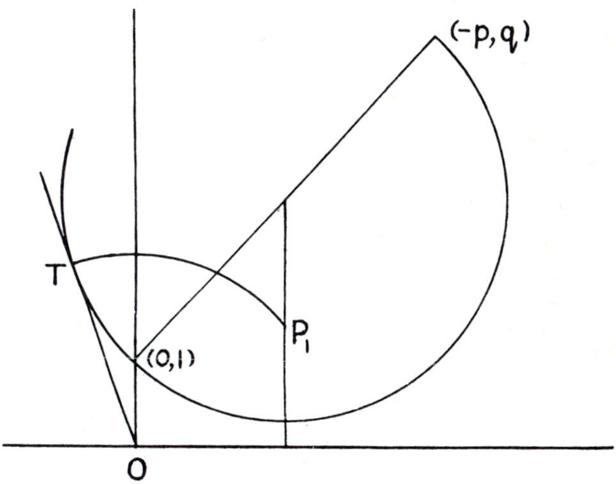

If the circle does not meet the x-axis, then Wigand replaces the given equation by

$$x^2+px+\tfrac{1}{2}p^2-q=0;$$

if the roots of this equation are $u \pm v$, then the roots of the original equation are $u \pm iv$.

† Wigand, K., *Praxis der Mathematik* (Cologne, June 1959).

Ingenious though Wigand's method is, it is preferable to have a construction in which the roots, whether real or imaginary, are found from the same circle. If the circle C on the line joining (0, 1) to $(-p, q)$ as diameter does not meet the x-axis, we draw OT the tangent to the circle from the origin O, and find the points of intersection P_1, P_2 of the circle centre O, radius OT, with the line through the centre of C parallel to the y-axis. If the coordinates of P_1, P_2 are $(u, \pm v)$ then the roots of the given quadratic equation are $u \pm iv$.

For $v^2 = OP_1{}^2 - u^2 = OT^2 - (\tfrac{1}{2}p)^2 = q - (\tfrac{1}{2}p)^2$ and $u = \tfrac{1}{2}p$.

The pupil who later proceeds to study coaxial systems of circles will have the fun of recognising in this construction the theorem that circles orthogonal to a coaxial system pass through the limiting points (point circles) of the system, and that if the circles of the system do *not* meet the radical axis in real points, then the circles of the orthogonal system have real intersections with the line of centres.

3.2 A Graphical Representation of Quadratic Equations

The following graphical representation of quadratic equations gives immediate information about the nature of the roots and can yield rough approximations to them.

Let the equation be

$$x^2 + ax + b = 0. \tag{1}$$

Take rectangular cartesian coordinates plotting a against b so that every point in the plane represents an equation with real coefficients. The locus of all equations having equal roots is then the parabola

$$a^2 = 4b. \tag{2}$$

All points inside the parabola represent equations with complex roots and all points outside equations with real roots. Moreover, all equations having one real root in common lie on a tangent to this parabola and the common root is given by $\sqrt{}$ (ordinate of the point of contact).

The condition that (1) should have one root α is

$$a\alpha + b + \alpha^2 = 0, \tag{3}$$

showing that the locus of equations with one common root is a straight line.

From (2) and (3), $a^2+4a\alpha+4\alpha^2 = 0,$

or $(a+2\alpha)^2 = 0;$

$$a = -2\alpha.$$

Thus (3) is a tangent to (2) at the point $(-2\alpha, \alpha^2)$.

3.3 A Nomogram for the Solution of Quadratic Equations

The roots of the quadratic equation

$$ax^2+bx+c = 0$$

are given by the intersection of the parabola

$$y = x^2$$

with the straight line

$$ay+bx+c = 0,$$

whose intercepts on the axes of x and y are $-c/b$, $-c/a$. This may be used as the basis of a nomogram as shown in Fig. 1, where the x-axis is labelled c/b, the y-axis labelled c/a (both with reversed signs) and the values of x are marked on the curve.

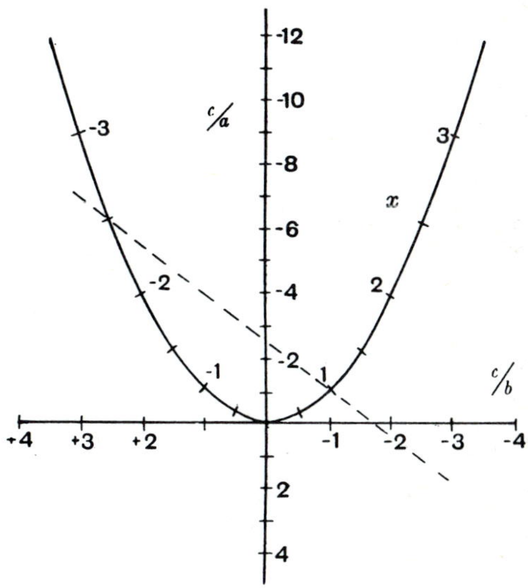

FIG. 1

Such a nomogram has very limited application, but the useful range can be increased by projecting the coordinate axes OX, OY and the line at infinity into the sides CB, CA, AB of an equilateral triangle, and the points $(1, 0)$, $(0, 1)$ into the mid-points of the sides BC, AC. The projection of the parabola is then a circle touching AB, BC at A and C. Using areal coordinates the equations of the curve and line become

$$\beta^2 - \alpha\gamma = 0$$
$$a\alpha + b\beta + c\gamma = 0.$$

The formerly linear scales of x and y are projected into proportional scales of γ/β, γ/α. For convenience the latter scale is reversed and graduated in terms of α/γ; AB is also graduated in terms of β/γ. A method of graduating such a scale has been described in the *Mathe-*

FIG. 2

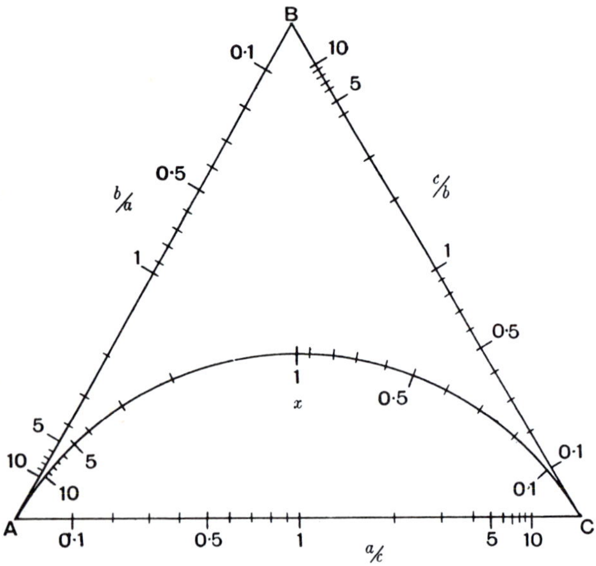

matical Gazette (XXXVII, p. 39). In the nomogram, shown in Fig. 2, these scales are labelled respectively c/b, a/c, b/a. The graduations on the curve, which is the locus $\gamma/\beta = \beta/\alpha$, are obtained by projecting either the c/b scale or the b/a scale from the opposite vertex of the triangle of reference.

To keep the size of the nomogram within reasonable bounds the sides are graduated for positive values only, and only that part of the curve within the triangle is drawn. For positive values of c the nomogram is used as indicated in Fig. 3(i). The straight-edge, laid between the appropriate points on the b/a, c/b scales (the sign of b being ignored), lies along the line

$$a\alpha - |b|\beta + c\gamma = 0.$$

The intersections of this line with the graduated arc give the numerical values of the roots. For negative values of c the two roots are found separately as indicated in Fig. 3(ii). Here the equations of the two lines along which the straight-edge is laid are

$$a\alpha + |b|\beta + c\gamma = 0, \quad a\alpha - |b|\beta + c\gamma = 0$$

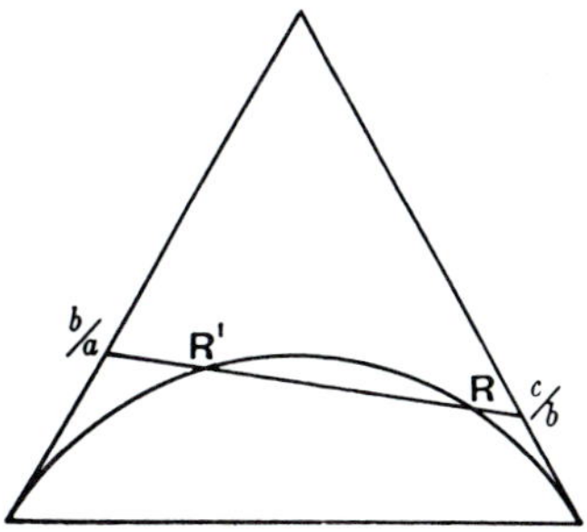

FIG. 3(i)
c positive. The roots (R, R') have the opposite sign to b.

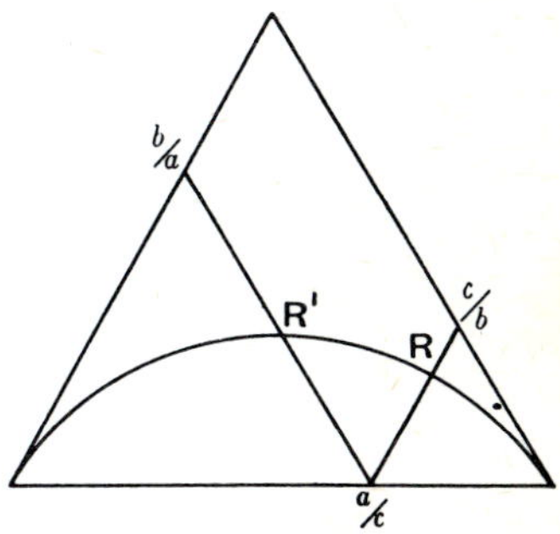

FIG. 3(ii)
c negative. The numerically smaller root has the sign of b; the numerically larger root has the opposite sign.

and again the intersections of these lines with the graduated arc give the numerical values of the roots.

Finally on quadratic equations we have an easy problem which leads on to a simple paradox.

3.4 An Interesting Equation

To find a quadratic equation

$$x^2+ax+b=0$$

whose two roots are the coefficients a, b.

Since a, b are two roots of $x^2+ax+b=0$

$$a+b = -a \qquad (1)$$

$$ab = b \qquad (2)$$

by the relation between the roots and the coefficients of a quadratic equation.

From (2) we have $b = 0$ or $a = 1$.
When $b = 0$, we have $a = 0$ from (1).
When $a = 1$, we have $b = -2$ from (1).
The case $a = b = 0$ is trivial.
Hence the required equation is

$$x^2+(1)x+(-2) = x^2+x-2 = 0.$$

It is natural enough to attempt to solve this problem from first principles, i.e. direct from the equations to be satisfied, avoiding any appeal to 'known' results. Strangely enough a disaster, or at least a paradox, ensues. The solutions of

$$a^2+a^2+b=0 \qquad (3)$$

$$b^2+ab+b=0 \qquad (4)$$

found by factorising (4) and substituting each alternative in (3), are

$$b=0,\ a=0 \quad \text{and} \quad a=1,\ b=-2 \quad \text{and}$$
$$a=-\tfrac{1}{2},\ b=-\tfrac{1}{2};$$

these are genuine solutions of equations (3) and (4), but the third pair of values does not satisfy the original problem.

Where is the flaw?

3.5 The Condition for a Cubic Equation to have Real Roots

The text-book proofs of the necessary and sufficient condition $4a^3 \geqq b^2$ for the cubic equation $x^3-3ax+b=0$ to have three real roots usually depend upon calculus methods, or upon calculating the products of the squares of the differences between the roots in pairs. The following proof is based upon:

(i) $b^2 \geqq 4ac$ is a necessary and sufficient condition for the roots of the quadratic equation $ax^2+bx+c = 0$ to be real;
(ii) every cubic equation with real coefficients has at least one real root, since unreal roots occur in conjugate pairs;
(iii) if all the roots are real and their sum is zero, then at least one is positive and at least one is negative.

We prove first that, if the equation $x^3-3ax+b = 0$ has three real roots, then $4a^3 \geqq b^2$. Define a real number λ as follows:

if $b > 0$, take λ to be the negative root;
if $b < 0$, take λ to be the positive root.

Note that in either case $-b/\lambda$ is positive. (If $b = 0$, the cubic is factorisable: one root is zero, and the others are real if $a \geqq 0$.) Then

$$x^3-3ax+b \equiv (x-\lambda)(x^2+\lambda x-b/\lambda),$$

where

$$3a = \lambda^2+b/\lambda \tag{1}$$

and, since the other roots are real,

$$\lambda^2 \geqq -4b/\lambda > 0. \tag{2}$$

Eliminating b between (1) and (2) we obtain

$$a \geqq \lambda^2/4.$$

Hence $a^3 \geqq \lambda^6/64$ and (from (2)) $\lambda^4/16 \geqq b^2/\lambda^2$, so that

$$4a^3 \geqq b^2. \tag{3}$$

To prove the converse (that if $4a^3 \geqq b^2$, then all the roots are real), let λ be any real root (known to exist, by (ii) above). We then have to deduce that $\lambda^2 \geqq -4b/\lambda$ from (1) and (3).

By eliminating a between (1) and (3), it can be shown that

$$(\lambda^3+4b)(2\lambda^3-b)^2/\lambda^3 \geqq 0;$$

and since $(2\lambda^3-b)^2$ and λ^2 are positive, $(\lambda^3+4b)/\lambda \geqq 0$, so that

$$\lambda^2 \geqq -4b/\lambda.$$

3.6 The Real Roots of a Cubic Equation

A uniform system for finding the real roots of $x^3+qx+r=0$ is as follows:

I. If $4q^3+27r^2 \leqq 0$, then $x=\lambda \sin u$ where $\lambda=\sqrt{(-4q/3)}$, $\sin 3u = -3r/q\lambda$.

II. If $4q^3+27r^2>0$, then (a) $q>0$:
$x=\lambda \sinh u$ where $\lambda=\sqrt{(4q/3)}$, $\sinh 3u = -3r/q\lambda$,
(b) $q<0, r<0$:
$x=\lambda \cosh u$ where $\lambda=\sqrt{(-4q/3)}$, $\cosh 3u = 3r/q\lambda$,
(c) $q<0, r>0$:
$x=-\lambda \cosh u$ where $\lambda=\sqrt{(-4q/3)}$, $\cosh 3u = -3r/q\lambda$.

Of course II does not give the complex roots, but once the real root is known the sum and product of the complex roots can be written down and they can be found from their quadratic.

3.7 Complex Roots of a Cubic Equation from a Graph

The aim of this note is to show how the complex roots of a cubic equation can be found from a graph.

Suppose we have drawn the graph of

$$y=f(x)=(x-a)(x-b+ic)(x-b-ic),$$

or

$$f(x)=(x-a)(x^2-2bx+b^2+c^2),$$

then the tangent at $B[b,(b-a)c^2]$ passes through $A\,(a,0)$.
For

$$f'(x)=x^2-2bx+b^2+c^2+2(x-a)(x-b)$$

and

$$f'(b)=c^2.$$

The equation of the tangent to the curve at B is

$$y-(b-a)c^2 = c^2(x-b)$$

which passes through $(a,0)$.

The equation $f(x)=0$ has roots $a, b\pm ic$.

Thus we have shown that the tangent to the curve from the real root (i.e., where curve cuts x-axis) touches the curve at a point, the x-coordinate of which is the real part of the complex roots. If this

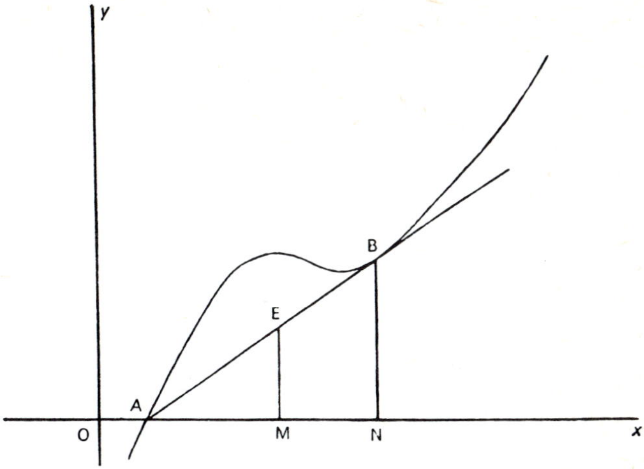

tangent is drawn by eye, the value of b will probably be as accurate as most graphical answers.

Since $\tan(\angle BAN) = c^2$, if we draw $EM \perp Ox$ with $AM = 1$ unit, then $EM = c^2$; c can then be found.

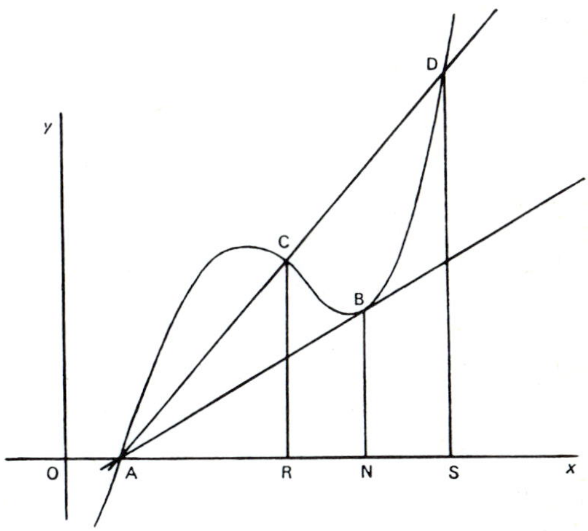

We can also find c by considering two other points on the cubic, namely

$$C[b-c, 2c^2(b-c-a)] \quad \text{and} \quad D[b+c, 2c^2(b+c-a)],$$

i.e., those points whose x-coordinates are $b-c$, $b+c$, which are the result of omitting i from the complex roots.

The line CD is

$$y = 2c^2(x-a)$$

which also passes through $A(a, 0)$ with gradient $2c^2$.

So if we draw a second line through A with gradient twice that of the tangent AB which cuts the curve at C and D, then the x-coordinates of B, C, D are b, $b-c$, $b+c$.

Therefore $RN = NS = c$ which is the imaginary part of the complex roots.

3.8 Graphs of Cubic Functions

In the teaching of graphs of cubic functions in schools, the symmetry of all cubic graphs about a point mid-way between the stationary values of the function does not appear to be specifically mentioned although it is a useful property to know when sketching graphs.

For, consider the general cubic:

$$y = ax^3 + bx^2 + cx + d.$$

By differentiation the mid-point of the line joining the stationary values is:

$$\left(\frac{-b}{3a}, \frac{2b^3}{27a^2} - \frac{cb}{3a} + d\right),$$

and transferring the origin to this point gives:

$$y = ax\left(x^2 - \frac{(b^2 - 3ac)}{3a^2}\right)$$

which is symmetrical about the new origin. If $b^2 < 3ac$ the stationary points are 'imaginary' and this is also the condition for not more than one real root of the equation $y = k$ to be possible; but the property of symmetry remains.

The result may be proved another way.

By shifting the origin a suitable distance along the x-axis, the equation of the curve may be expressed in the form

$$y = Ax^3 + Bx + C.$$

A shift of the origin along the y-axis enables us to express the equation in the form

$$y = Ax^3 + Bx.$$

The curve is symmetrical about the new origin, and so are the two stationary points.

The following property is also interesting.

If a graph of a cubic function $y = f(x)$ meets the x-axis in the points $(a, 0)$, $(b, 0)$, $(c, 0)$, then the tangent at the point

$$\left(\frac{a+b}{2}, f\left(\frac{a+b}{2}\right)\right)$$

passes through $(c, 0)$.

Let

$$y = f(x) \equiv k(x-a)(x-b)(x-c),$$

then we have

$$f\left(\frac{a+b}{2}\right) = -k(a-b)^2(a+b-2c)/8,$$

$$f'(x) = k(3x^2 - 2(a+b+c)x + (bc+ca+ab)),$$

and therefore

$$f'\left(\frac{a+b}{2}\right) = -k(a-b)^2/4.$$

So that the equation of the tangent is

$$y + k(a-b)^2(a+b-2c)/8 = -k(a-b)^2\left(x - \frac{a+b}{2}\right)\bigg/4$$

and $x = c$, $y = 0$ satisfy the equation.

The property can be proved without the use of calculus as follows:

If the cubic is given by $y = k(x-a)(x-b)(x-c)$, and any line through $(c, 0)$ is taken to be $y = m(x-c)$, where m can vary, then the line meets the curve where

$$k(x-a)(x-b)(x-c) = m(x-c).$$

$\therefore$ apart from $x = c$, the x-values of the points of intersection are given by

$$x^2 - (a+b)x = \frac{m}{k} - ab.$$

These values coincide if

$$x = \frac{a+b}{2} \quad \text{and} \quad -\frac{(a+b)^2}{4} = \frac{m}{k} - ab,$$

i.e.

$$m = -\frac{k}{4}(a-b)^2.$$

Thus the tangent at $x = \dfrac{a+b}{2}$ passes through $(c, 0)$ and has gradient $-\dfrac{k}{4}(a-b)^2$.

The result may also be proved in this way.

By shifting the origin a suitable distance along the x-axis, the equation of the cubic curve may be expressed in the form

$$y = Ax^3 + Bx + C.$$

Lemma

Any line (not parallel to the y-axis) meets the curve in three points (two of which may be imaginary) for which the sum of the x-coordinates is zero, as we see by substituting $y = mx + n$.

From the lemma, we see that for the three points where the curve meets the x-axis, we have $a+b+c = 0$. Using the lemma again, we see that the tangent at the point for which $x = \frac{1}{2}(a+b) = -\frac{1}{2}c$ meets the curve again at the point for which $x = c$.

Since the proof is independent of C, the result is true for any line parallel to the x-axis.

3.9 The Remainder Theorem

The usual school demonstration of the Remainder Theorem consists of either (1) performing the actual 'long division' of some typical polynomial such as $ax^3 + bx^2 + cx + d$ by $x - \alpha$, a device which laboriously churns out a quotient which is of no immediate interest; or (2) more abstractly, asserting the identity $P(x) \equiv (x-\alpha) \,.\, Q(x) + R$, and then setting x equal to α, a step which causes uneasiness in the minds of pupils (and teachers) if they get as far as (and no further than) questioning the validity of division by a divisor which becomes zero as soon as its work is done.

Here is another demonstration.

We recall that to divide is to express a number as a multiple of the divisor. In a simple case like $45 \div 7$ it amounts to rewriting '4 tens +5' as '6 sevens+3'. More generally, to divide the polynomial $f(x) \equiv a+bx+cx^2+dx^3+\ldots$ by $x-\alpha$ is to rewrite it as a sum of multiples of powers of $x-\alpha$. To accomplish this transformation, set $x-\alpha = y$, i.e. $x = \alpha+y$. Hence

$$f(x) \equiv a+b(\alpha+y)+c(\alpha+y)^2+d(\alpha+y)^3+\ldots.$$

The first term of each bracket yields in sum an absolute term

$$a+b\alpha+c\alpha^2+d\alpha^3+\ldots \equiv f(\alpha).$$

The other terms in the expansions all contain y as a factor, and so in sum may be written as $y \,.\, Q = (x-\alpha) \,.\, Q$, where Q then is the quotient on division of $f(x)$ by $x-\alpha$

i.e.
$$f(x) \equiv (x-\alpha) \,.\, Q+f(\alpha);$$

and the remainder term $f(x)$ has been found without either performing a long division or taking the step—now seen to be an irrelevance —of setting $x = \alpha$.

If doubt still exists in some minds about the validity of the usual school proof of the Remainder Theorem, the following may be of help.

Let P_r be a polynomial of degree r in x; $r = 0, 1, 2, \ldots$. So P_4 will denote an expression of type $cx^4+dx^3+ex^2+fx+g$, P_1 one of type $ax+b$ and P_0 a constant—where $a, b, c, \ldots$ are independent of x. In the so-called division process below, we suppose only $a, c \neq 0$.

$P_1)\ P_4 \qquad\qquad (q_1+q_2+q_3+q_4$

$\underline{q_1 P_1}$

$P_4 - q_1 P_1$

$\underline{q_2 P_1}$

$P_4-(q_1+q_2)P_1$

$\underline{q_3 P_1}$

$P_4-(q_1+q_2+q_3)P_1$

$\underline{q_4 P_1}$

$P_4-(q_1+q_2+q_3+q_4)P_1 = P_0$, the remainder free from x.

Hence $P_4 \equiv (q_1+q_2+q_3+q_4)P_1+P_0$.

To find q_1, cx^4 is divided by ax; but the equality $\dfrac{cx^4}{ax} = \dfrac{c}{a}x^3$ or q_1 is not used: it would forbid x to have the value 0. Instead it is the multiplication version, $cx^4 = \dfrac{c}{a}x^3 \,.\, ax$ which is employed in the process: this holds for *all* values of x. Also, the first subtraction holds for *all* values of x, yielding as remainder $P_4 - q_1P_1$, a polynomial of degree 3 at most. The rest of the division, until the remainder P_0 free from x is reached, is but a repetition of the processes already described. What has been established is that corresponding to given polynomials in x P_1, P_4 a polynomial $q_1+q_2+q_3+q_4$ or P_3, say, and a constant P_0 *exist* such that

$$P_4 \equiv P_1 \,.\, P_3 + P_0 \tag{1}$$

where x is totally unrestricted. This is what the division process primarily shows beyond doubt. The common misapprehension surrounding the usual proof of the Remainder Theorem centres on the illusion that the identity just written derives from the almost equivalent one

$$\frac{P_4}{P_1} \equiv P_3 + \frac{P_0}{P_1} \tag{2}$$

Instead, this is derived from (1), with the limitation that x must not make P_1 vanish!

Knowing that an identity such as (1) *exists*, we can without division by the method of undetermined coefficients find the forms of P_3, P_0 completely by substituting $(4+1)$ values of x in the assumed form. Usually, however, only P_0 is required, and one substitution suffices to fix this (in ignorance of P_3): it is the value that makes P_1 vanish. Hence P_0, the remainder is the result of substituting in the dividend P_4 that value of x, namely $-\dfrac{b}{a}$, which makes P_1 vanish. Similar remarks apply when the dividend is P_r instead of P_4.

The 'division' process likewise leads to the result that corresponding to given polynomials P_s, $P_r (s \leqq r)$, other polynomials P_{r-s}, P_{s-1} exist such that

$$P_r \equiv P_s P_{r-s} + P_{s-1} \tag{3}$$

for all x. Of course the coefficients (one or more) in P_{r-s} and P_{s-1} may vanish (other than the leading one in P_{r-s}). In any case these coefficients $(r-s+1+s-1+1)$ or $(r+1)$ in number can be found by the method of undetermined coefficients by $(r+1)$ substitutions at most. Usually only those of P_{s-1} are required and these are found by substituting the zeros of P_s in (3) or, in the case of repeated roots, in (3) and in the equalities found from (3) by repeated differentiation with respect to x.

3.10 The Inequality between the Arithmetic and Geometric Means

There must be more than a dozen proofs of this result. Here we offer just three.

Let A_n be the Arithmetic mean of $a_1, a_2, \dots a_n$, and G_n the Geometric mean.

For positive values of x write

$$f(x) \equiv \frac{nA_n + x}{n+1} - \{G_n{}^n x\}^{1/n+1}.$$

Then

$$f'(x) = \frac{1}{n+1}\left\{1 - \left(\frac{G_n}{x}\right)^{n/n+1}\right\}.$$

Hence $f'(x) = 0$, and $f(x)$ has a *minimum* value when $x = G_n$.

Also

$$f(G_n) = \frac{n}{n+1}\{A_n - G_n\}.$$

Now *assume*

$$A_n > G_n.$$

Then

$$A_{n+1} - G_{n+1} = f(a_{n+1}) \geqq f(G_n) > 0,$$

and the theorem follows by induction from $A_2 > G_2$.

Moreover it is clear that equality is possible only when all the quantities are equal.

If the use of the calculus in an algebraical theorem is objected to, we may proceed a little differently:

$$f(x) - f(G_n) = \frac{1}{n+1}\{x - G_n\} - G_n{}^{n/n+1}\{x^{1/n+1} - G_n{}^{1/n+1}\}.$$

Put $x = y^{n+1}$, $G_n = g^{n+1}$,

$$f(x)-f(G_n)=\frac{1}{n+1}\{y-g\}\{y^n+y^{n-1}g+\ldots+g^n\}-g^n\{y-g\}$$

$$=(y-g)\left\{\frac{y^n+y^{n-1}g+\ldots+g^n}{n+1}-g^n\right\}.$$

If $y > g$, both factors are positive; if $y < g$, both are negative. Hence $f(x) > f(G_n)$ as before.

The third proof is as follows:

Let P denote the product, and S the sum of n positive numbers $a_1, a_2, \ldots, a_n$. The inequality

$$\sqrt[n]{P} \leqq S/n \tag{1}$$

is true for $n = 2$. Let $a > 0$,

$$Q = (Pa)^{1/n(n+1)},$$

and suppose that (1) holds for some $n \geqq 2$. Then three applications of (1) lead to

$$Q^n = \{(Qa^{(n-1)/n})(P^{1/n})^{n-1}\}^{1/n}$$

$$\leqq \frac{1}{n}(Qa^{(n-1)/n})+\frac{n-1}{n}P^{1/n}$$

$$\leqq \frac{1}{n}(Q^n a^{n-1})^{1/n}+\frac{n-1}{n^2}S$$

$$\leqq \frac{1}{n}\left(\frac{1}{n}Q^n+\frac{n-1}{n}a\right)+\frac{n-1}{n^2}S.$$

Since $n-1 > 0$, the extreme members give

$$\sqrt[n+1]{(Pa)} \leqq (S+a)/(n+1).$$

Thus (1) holds for $n+1$ numbers, and hence generally. The proof also shows that there is equality in (1) if, and only if, $a_1 = a_2 = \ldots = a_n$.

3.11 Applications of the Inequality of the Means

Two examples of applications of the well-known inequality between the arithmetical and geometrical means.

I To prove $\lim_{n\to\infty} n^{1/n} = 1$.

We have

$$1 < n^{1/2n} = \{n^{n/2}\}^{1/n^2} < \{1+1+\ldots+1+\sqrt{n}+\sqrt{n}+\ldots+\sqrt{n}\}/n^2,$$

the numerator containing n^2-n units and $\sqrt{n}$ repeated n times, and so

$$1 < n^{1/2n} < 1-1/n+1/\sqrt{n} < 1+1/\sqrt{n},$$

whence

$$1 < n^{1/n} < 1+3/\sqrt{n},$$

and the result follows.

II To prove $(1+1/n)^n$ is an increasing function of n.

We have

$$\left\{\left(1+\frac{1}{n}\right)^n\right\}^{1/(n+1)} < \left\{1+\left(1+\frac{1}{n}\right)+\left(1+\frac{1}{n}\right)+\ldots+\left(1+\frac{1}{n}\right)\right\}\Big/(n+1)$$

$$= \frac{1+(n+1)}{n+1} = 1+\frac{1}{n+1},$$

whence

$$\left(1+\frac{1}{n}\right)^n < \left(1+\frac{1}{n+1}\right)^{n+1}$$

3.12 The Binomial Theorem for any Index

The following proof is possibly a little shorter than several existing ones: it uses the theorem on differentiation of a power series.

Let $$m_r = m(m-1)\ldots(m-r+1)/r!,\ m_0 = 1.$$

Then $$(m+1)_r - m_r = m_{r-1} \quad (r \geqq 1). \tag{1}$$

If

$$f(m) = \sum_0^\infty m_r x^r \quad (|x| < 1), \tag{2}$$

then

$$f(m+1)-f(m) = \sum_1^\infty \{(m+1)_r - m_r\}x^r$$

$$= x\sum_1^\infty m_{r-1}x^{r-1}, \quad \text{by (1)}$$

$$= xf(m), \quad \text{by (2).}$$

Hence

$$\frac{f(m+1)}{(1+x)^{m+1}} = \frac{f(m)}{(1+x)^m} = \frac{f(m-1)}{(1+x)^{m-1}} = \dots .$$

Thus
$$f(m) = \phi(x)(1+x)^m, \tag{3}$$

where $\phi(x)$ is independent of m. But

$$df(m)/dx = \sum_1^\infty m_r r x^{r-1} = m\sum_0^\infty (m-1)_r x^r$$

$$= mf(m-1), \quad \text{by (2).} \tag{4}$$

That is, by (3) and (4),

$$(1+x)^m\phi'(x) + m(1+x)^{m-1}\phi(x) = m(1+x)^{m-1}\phi(x)$$

and hence

$$\phi'(x) = 0, \quad \phi(x) = K, \text{ a constant.}$$

Thus
$$f(m) = K(1+x)^m$$

and
$$K = f(0) = 1;$$

that is,
$$f(m) = (1+x)^m.$$

3.13 Pascal's Triangle

On first introducing Pascal's triangle to a form, give them some while to look for patterns within this greatest of all patterns by themselves. With the minimum of help everybody finds something; and some find quite abstruse patterns. Listed below are some of the odder ones.

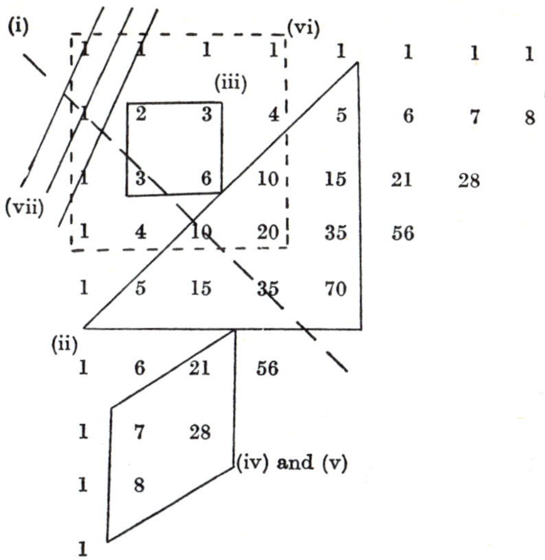

(i) Each number in the marked diagonal is the sum of all those above it.

(ii) All numbers within triangles situated 'like the one shown', stretching from the second row to the second column, are divisible by the number at the top iff this is prime.

(iii) Any 2 by 2 determinant in the second and third column is equal to its top right number.

(iv) A parallelogram like this in the second and third columns has its 'determinant' equal to its bottom right number.

(v) If you add up the numbers in such a parallelogram, the sum is the square of the bottom left number.

(vi) A determinant of any size starting from the top left corner is equal to 1.

(vii) Adding along the diagonals of which the first three are shown gives, in succession, the terms 1, 1, 2, 3, 5, 8, of a Fibonacci series.

None of these is particularly difficult to prove; but most of them are interesting for the new light they shed on old ideas. (v), for instance, is just a re-presentation of the old idea that a square is the sum of two consecutive triangular numbers.

Pascal's triangle can be extended to give the coefficients of the expansion of $(1+x)^{-n}$, as follows:

$(1+x)^{-5}$	-	1	−5	15	−35	70	−126	210	−330	...
$(1+x)^{-4}$	-	1	−4	10	−20	35	−56	84	−120	...
$(1+x)^{-3}$	-	1	−3	6	−10	15	−21	28	−36	...
$(1+x)^{-2}$	-	1	−2	3	−4	5	−6	7	−8	...
$(1+x)^{-1}$	-	1	−1	1	−1	1	−1	1	−1	—
$(1+x)^{0}$	-	1	0	0	0	0	0	0	0	...
$(1+x)^{1}$	-	1	1	0	0	0	0	0	0	...
$(1+x)^{2}$	-	1	2	1	0	0	0	0	0	...
$(1+x)^{3}$	-	1	3	3	1	0	0	0	0	...
$(1+x)^{4}$	-	1	4	6	4	1	0	0	0	...
. . .	.	.	.	.	.	.	.	.	.	.

To extend the table upwards, all we have to note is that any entry is equal to the entry below minus the entry to the left of the required entry.

The following is another way of looking at the same result.

Pascal's triangle has long been used to give the coefficients in the expansion of the Binomial theorem for positive integral indices. But if we construct a Pascal triangle:

1					
1	1				
1	2	1			
1	3	3	1		
1	4	6	4	1	
1	5	10	10	5	1
1	6	15	20	...	

it will be seen that the vertical columns give the coefficients for the expansion of the Binomial theorem for negative integral indices.

The idea of extending its application leads to the following method of calculating the coefficients in a trinomial expansion, say $(a+b+c)^4$. (See opposite).

The diagram represents a model constructed of equilateral card-board triangles joined at the vertices by strips of balsa wood to form a regular tetrahedron. Each number is the sum of its three nearest neighbours in the layer above: for instance, the ringed 12 is the sum of

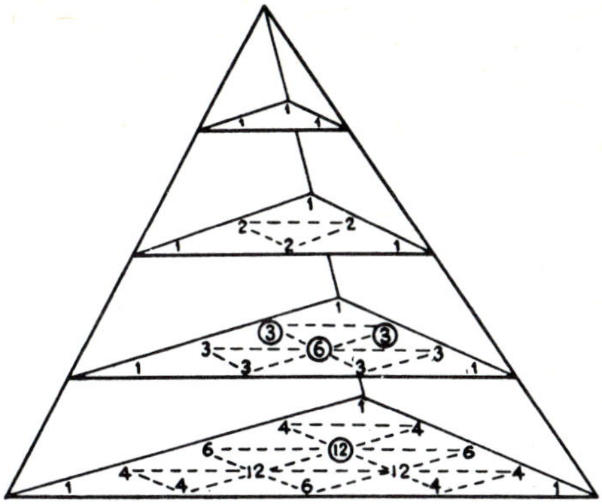

the ringed 6 and 3's. The numbers in the nth layer give the coefficients in the expansion of $(a+b+c)^n$, allocated according to the scheme

$$
\begin{array}{cccccccc}
 & & & a^n & & & & \\
 & & a^{n-1}b & & a^{n-1}c & & & \\
 & a^{n-2}b^2 & & a^{n-2}bc & & a^{n-2}c^2 & & \\
a^{n-3}b^3 & & a^{n-3}b^2c & & a^{n-3}bc^2 & & a^{n-3}c^3 & \\
\cdots & & & & & & & \cdots \\
\cdots & & & & & & & \cdots \\
b^n \quad b^{n-1}c \quad b^{n-2}c^2 & & \cdots & & \cdots & & b^2c^{n-2} \quad bc^{n-1} \quad c^n &
\end{array}
$$

The method depends, of course, on the identity

$$\frac{n!}{p!q!r!} = \frac{(n-1)!}{(p-1)!q!r!} + \frac{(n-1)!}{p!(q-1)!r!} + \frac{(n-1)!}{p!q!(r-1)!}$$

where $p+q+r = n$.

3.14 Sums of Powers of the Natural Numbers: Proofs by Dissection Methods

Simple Methods for $S_1, S_2, S_3 \left(S_p = \sum_1^n r^p \right)$

I Consideration of the oblique rows of squares on a chess-board suggests the identity:

$$n^2 = 1+2+3+\ldots+(n-1)+n+(n-1)+\ldots+2+1. \qquad (1)$$

Adding n to each side

$$n^2+n = 2S_1 \quad \text{or} \quad S_1 = \tfrac{1}{2}(n^2+n).$$

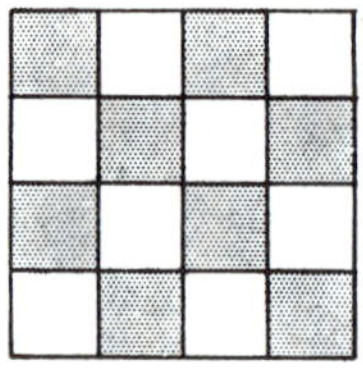

FIG. 1

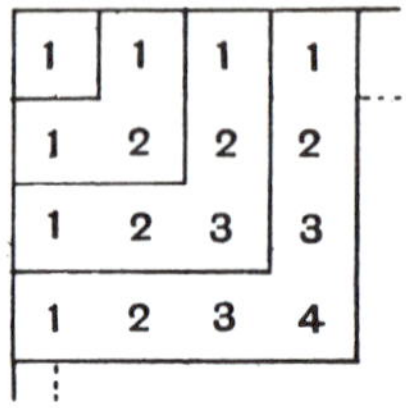

FIG. 2

II The sum of the numbers in this array (Fig. 2) is S_2, since the sum in the rth 'gnomon' is r^2.

But the sum in the rth row is

$$(1+2+\ldots+r)+(n-r)r = \tfrac{1}{2}r^2+\tfrac{1}{2}r+nr-r^2 = (n+\tfrac{1}{2})r-\tfrac{1}{2}r^2.$$

Thus adding the rows,

$$S_2 = (n+\tfrac{1}{2})S_1-\tfrac{1}{2}S_2,$$

or

$$S_2 = \frac{2n+1}{3}\,.\,S_1.$$

III The sum of the numbers in this array (Fig. 3) is S_3, since the sum in the rth 'gnomon' is

$$r+2r+3r+\ldots+(r-1)r+r^2+\ldots+2r+r = r^3 \text{ from (1).}$$

But in the rth row is rS_1.

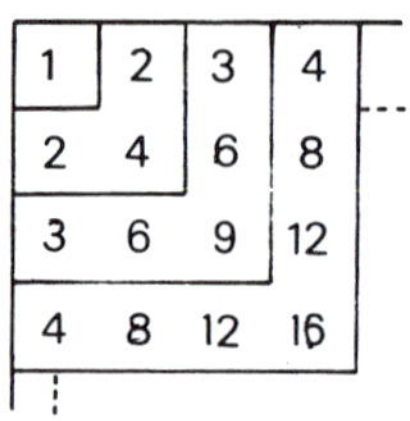

FIG. 3

Thus the sum by rows is $S_1{}^2$. Hence $S_3 = S_1{}^2$.

IV To show diagrammatically the relation $3S_2 = (2n+1)S_1$. (Fig. 4).

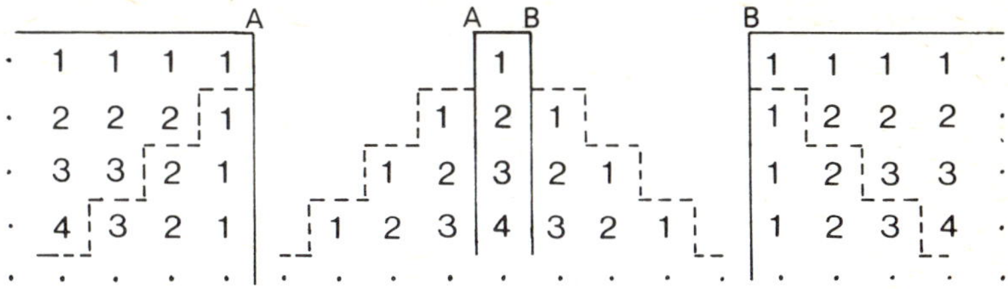

FIG. 4

The sum of the numbers in each of the three arrays is S_2, by considering the 'gnomons' of the squares and the rows of the central figure. Now slide the squares over the central figure until the dotted lines coincide.

The result is a rectangular array whose sum is $(2n+1)S_1$.

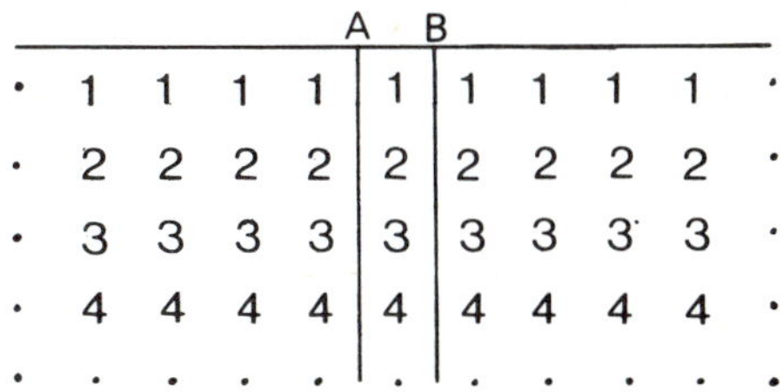

FIG. 5

Another demonstration of the use of gnomons to find the sum of the cubes of the natural numbers is as follows:

In any square of cells the addition of the number of cells taken by diagonals is $1+2+3+\ldots+(n-1)+n+(n-1)+\ldots+3+2+1$, and this is equal to n^2.

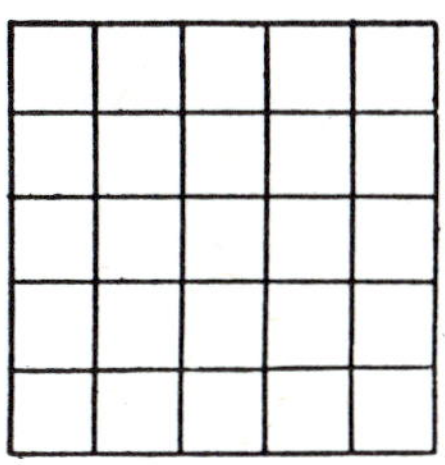

FIG. 6

Now consider a square of cells divided up so that there are 1, 2, 3, 4 . . . cells along the top edge and down the left-hand edge (Fig. 7).

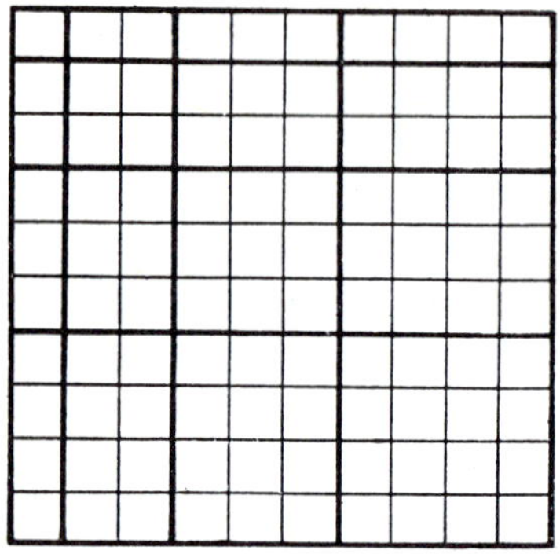

FIG. 7

The top left-hand cell counts 1. The gnomon of width 2 units contains $2 . 1+2 . 2+2 . 1$ cells (starting at the top, going down to the square (2 . 2) and then to the left).

The gnomon of width 3 units contains

$$3 . 1+3 . 2+3 . 3+3 . 2+3 . 1 \text{ cells;}$$

and the gnomon of width n units contains

$$n(1+2+3+\ldots+n+\ldots+3+2+1)$$

which by the above

$$= n . n^2 = n^3.$$

The complete square then contains

$$1+2^3+3^3+\ldots+n^3$$

cells; but the edge is

$$1+2+3+\ldots+n$$

units long.

Hence $$\sum n^3 = (\sum n)^2$$

or, alternatively:

Suppose Fig. 7 represents a layer of small cubes. Each gnomon can be built up into a solid cube. Thus: take the bottom right-hand square as the base, on it place the bottom strip to the left of it and the top strip above it; together they form an equal layer; the strip from the bottom and the strip from the right-hand side next to these form another equal layer, and so on. And a cube of n^3 small cubes is formed.

Fig. 8 makes this dissection by gnomons clearer.

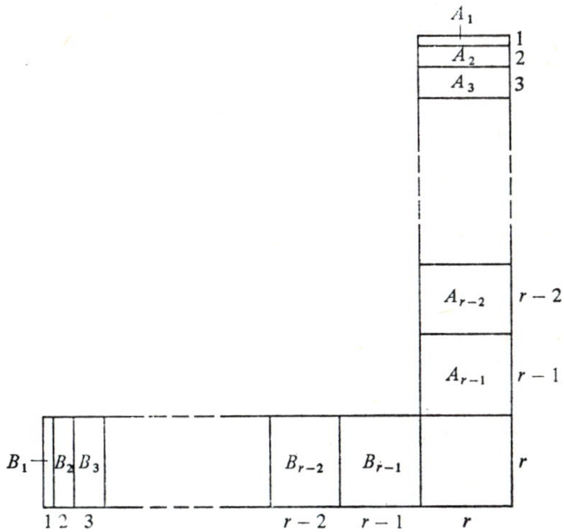

FIG. 8

A square of side $1+2+\ldots n$ can be divided into a unit square and $(n-1)$ gnomons of the type shown in the figure ($r = 2, 3, \ldots n$). The rectangles A_1 and B_{r-1} together make a square of side r; so do A_2 and B_{r-2}; and so on. Hence the whole area of the gnomon is $r \times r^2$ and that of the original square is $1^3+2^3+\ldots n^3$.

Here is an alternative method of showing diagrammatically that

$$3S_2 = (2n+1)S_1, \quad \text{where} \quad S_p = \sum_{r=1}^{n} r^p$$

Taking $n = 4$, the sum of the squares is equal to the sum of the numbers in the triangle of Fig. 9.

Now turn the triangle through 120° so that B becomes the vertex and CA the base, and then repeat the operation so that C becomes the vertex and AB the base; we thus get the two triangles of Fig. 10.

FIG. 9

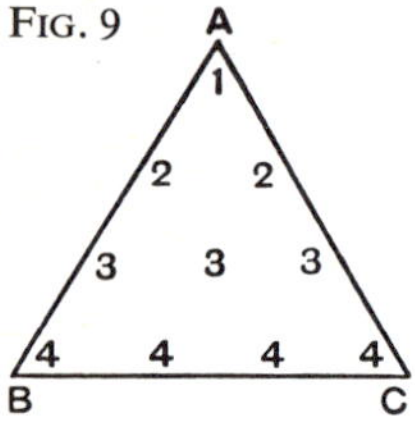

FIG. 10

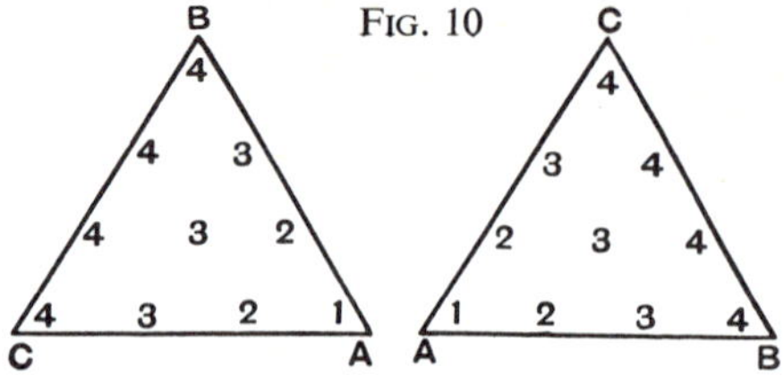

Now pile the triangles vertically one on top of the other, and add; in other words, construct a new triangle XYZ, in which each element is the sum of the corresponding elements in the three triangles ABC, BCA, CAB; this gives the triangle of Fig. 11, which equals $9(1+2+3+4)$.

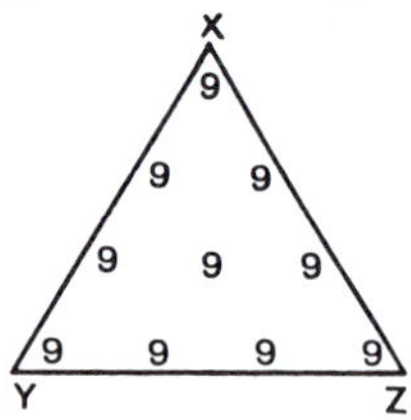

FIG. 11

It is easy to see that in the general case the elements of the triangle XYZ are all $2n+1$.

3.15 Sums of Powers of the Natural Numbers: Proofs by Algebraic Methods †

I The usual elementary method of obtaining the formula for $S_p(n)$, the sum of the pth powers of the first n natural numbers, is by 'differencing' some polynomial function of n, generally n^{p+1} or perhaps $n(n+1)\dots(n+p)$. We can get the formulae more rapidly if we difference $\{n(n+1)\}^r$ for the sums of odd powers S_{2r-1} and $(2n+1)\{n(n+1)\}^r$ for the sums of even powers S_{2r}. Thus

$$\{n(n+1)\}^r-\{n(n-1)\}^r = 2n^r\{{}^rC_1 n^{r-1}+{}^rC_3 n^{r-3}+\dots\}.$$

Summing from 1 to n we get

$$\{n(n+1)\}^r = 2\{{}^rC_1 S_{2r-1}+{}^rC_3 S_{2r-3}+\dots\}, \qquad (1)$$

the sum on the right ending with $2S_r$ or $2rS_{r+1}$ according as r is odd or even. Evidently $r = 1, 2$ give the elementary formulae

$$S_1 = \tfrac{1}{2}n(n+1), \quad S_3 = \tfrac{1}{4}n^2(n+1)^2.$$

† Note 1541 has been abbreviated. See also Tahta, D. G., Note 2753, *Mathematical Gazette*, Vol. 42, No. 339, p. 51.

Putting $r = 3, 4, 5$, say, and writing $N \equiv n(n+1)$ for brevity, we have

$$N^3 = 6S_5 + 2S_3, \qquad N^4 = 8S_7 + 8S_5,$$
$$N^5 = 10S_9 + 20S_7 + 2S_5,$$

which give without difficulty

$$S_5 = \tfrac{1}{6}N^3 - \tfrac{1}{12}N^2, \qquad S_7 = \tfrac{1}{8}N^4 - \tfrac{1}{6}N^3 + \tfrac{1}{12}N^2,$$
$$S_9 = \tfrac{1}{10}N^5 - \tfrac{1}{4}N^4 + \tfrac{3}{10}N^3 - \tfrac{3}{20}N^2.$$

The general formula begins

$$S_{2r+1} = \frac{N^{r+1}}{2(r+1)} - \frac{r-1}{12}N^r + \frac{r(r-2)(7r-1)}{720}N^{r-1} - \ldots,$$

but the form of the coefficients rapidly becomes obscure. We should note, however, that every S_{2r-1}, except of course S_1, has the factor N^2.

For the even powers we have

$$(2n+1)\{n(n+1)\}^r - (2n-1)\{n(n-1)\}^r$$
$$= 2n^r\left\{(2r+1)n^{2r} + \frac{2r-1}{3}\,{}^rC_2 n^{2r-2} + \ldots\right\},$$

and, on summation,

$$(2n+1)\{n(n+1)\}^r = 2\left\{(2r+1)S_{2r} + \frac{2r-1}{3}\,{}^rC_2 S_{2r-2} + \ldots\right\}, \qquad (2)$$

the last term being $2(r+2)S_{r+1}$ or $2S_r$ according as r is odd or even.

Evidently $r = 1$ gives the elementary result

$$S_2 = \tfrac{1}{6}n(n+1)(2n+1).$$

More generally, taking $r = 2, 3, 4$, say, we get

$$(2n+1)N^2 = 10S_4 + 2S_2, \qquad (2n+1)N^3 = 14S_6 + 10S_4,$$
$$(2n+1)N^4 = 18S_8 + 28S_6 + 2S_4,$$

and so

$$S_4 = (2n+1)(\tfrac{1}{10}N^2 - \tfrac{1}{30}N), \qquad S_6 = (2n+1)(\tfrac{1}{14}N^3 - \tfrac{1}{14}N^2 + \tfrac{1}{42}N),$$
$$S_8 = (2n+1)(\tfrac{1}{18}N^4 - \tfrac{1}{9}N^3 + \tfrac{1}{10}N^2 - \tfrac{1}{30}N).$$

The general formula begins

$$S_{2r} = \frac{2n+1}{2r+1}\left\{\frac{N^r}{2} - \frac{r(r-1)N^{r-1}}{12} + \frac{r(r-1)(r-2)(7r-1)}{720}N^{r-2} - \ldots\right\}.$$

Here S_{2r} always contains the factor $(2n+1)N$.

Comparison of the formulae for S_{2r}, S_{2r+1} suggests that, if in these formulae we replace the integer n by a variable x, we shall have generally

$$(2r+1)S_{2r}(x) = \frac{d}{dx}S_{2r+1}(x).$$

This can be established inductively by comparing the fundamental formulae (1), (2).

There are however more elementary methods of obtaining some of these results.

II Most students are somewhat surprised when $\sum_1^n r^3$ turns out to be $\frac{1}{4}n^2(n+1)^2 = \left(\sum_1^n r\right)^2$. The following simple, direct demonstration may add interest to this result. It can, of course be used as an alternative method of *proving* the formula for $\sum r^3$, if so desired.

Let $S_2(r)$ $(r \geqq 2)$ be the sum of all the possible products formed two at a time from the first r integers.

[Thus $S_2(4) = 1 \,.\, 2+1 \,.\, 3+2 \,.\, 3+1 \,.\, 4+2 \,.\, 4+3 \,.\, 4$.]

Clearly, (if $r \geqq 3$),

$$S_2(r) - S_2(r-1) = r\sum_1^{r-1} k$$

$$= \tfrac{1}{2}(r^3 - r^2). \qquad (1)$$

Define $S_2(0) = S_2(1) = 0$; then (1) holds for $r \geqq 1$.

Summing (1), then from $r = 1$ to n,

$$2S_2(n) = \sum_1^n r^3 - \sum_1^n r^2.$$

But obviously,

$$2S_2(n) = \left(\sum_1^n r\right)^2 - \sum_1^n r^2.$$

Hence the result.

III Alternatively, we have the following:
Let

$$u_n = \sum_{r=1}^{n} r = \tfrac{1}{2}n(n+1).$$

Then

$$u_n - u_{n-1} = n,$$

and

$$u_n + u_{n-1} = n^2.$$

Hence, by multiplication

$$u_n^2 - u_{n-1}^2 = n^3, \qquad (1)$$

and, consequently, by summation, it follows that

$$u_n^2 = \sum_{r=1}^{n} r^3,$$

i.e.

$$\left(\sum_{r=1}^{n} r\right)^2 = \sum_{r=1}^{n} r^3.$$

Several other interesting results can be obtained in this simple way. For example, since

$$u_n^2 + u_{n-1}^2 = \tfrac{1}{2}n^2(n^2+1), \qquad (2)$$

it follows, by multiplying (1) and (2) and summing, that

$$u_n^4 = \tfrac{1}{2}\sum_{r=1}^{n} (r^7 + r^5),$$

i.e.

$$2\left(\sum_{r=1}^{n} r\right)^4 = \sum_{r=1}^{n} r^7 + \sum_{r=1}^{n} r^5.$$

Similarly, we can easily deduce that

$$u_n^3 - u_{n-1}^3 = \tfrac{1}{4}n^3(3n^2+1),$$

and hence it follows that

$$4\left(\sum_{r=1}^{n} r\right)^3 = 3\sum_{r=1}^{n} r^5 + \sum_{r=1}^{n} r^3.$$

3.16 A Single-Scale Nomogram

Since the equation $x^3+ax+b=0$ has zero for the sum of its roots, the x-coordinates of the three intersections of the line $y=mx+c$ and the curve $y=x^3+px+q$ add to zero. Thus a curve $y=x^3+px+q$, graduated according to its x-coordinates, provides a single-scale nomogram for simple addition and subtraction. Graduated according to the antilogarithms of its x-coordinates, it becomes a nomogram for multiplication and division. In fact, appropriate graduation will give a nomogram for performing any operation which may be executed by a suitably graduated slide rule.

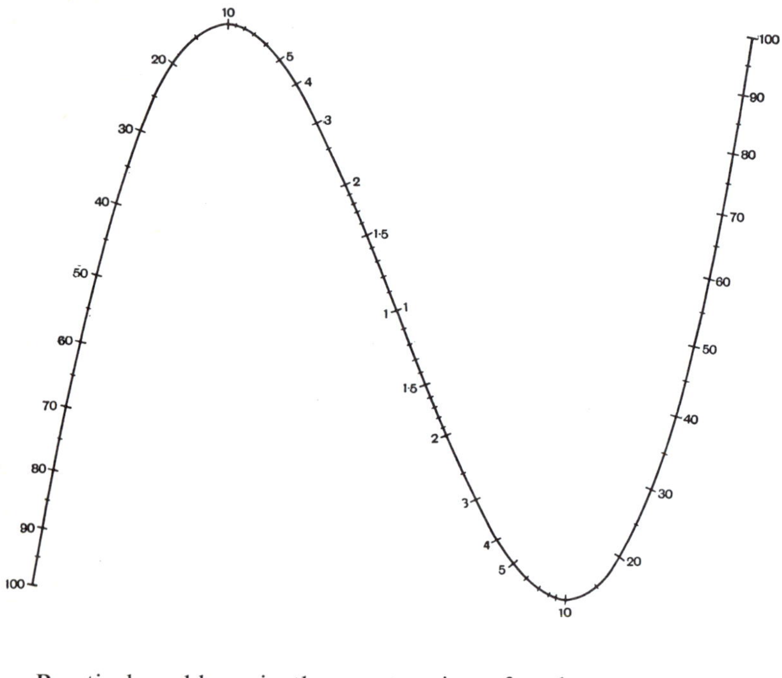

Practical problems in the construction of such a nomogram are (a) selection of a suitable cubic curve, e.g. in order to give a good 'cut' in the majority of cases, (b) selection of suitable scales for the x- and y-axes, (c) selection of the range of x-values to be used, (d) selection of the range of graduations to be displayed on the curve, e.g. the base of the antilogarithms in the example given above.

The diagram is for multiplication and division. For the former, join two numbers on the same half of the curve by a straight line, and read its third intersection with the (other half of the) curve. For division, connect numbers on opposite halves of the curve.

3.17 The Sign of a Rational Function

Consider the inequality $\dfrac{(x-3)(x+2)}{(x-1)(x+4)} < 0$. Some textbooks on elementary mathematics proceed to go into a trauma of tables from $-\infty$ to $+\infty$ to show that $-4 < x < -2$ and $1 < x < 3$, but if the numerator and denominator are sketched graphically as shown, the answer can be written down at once.

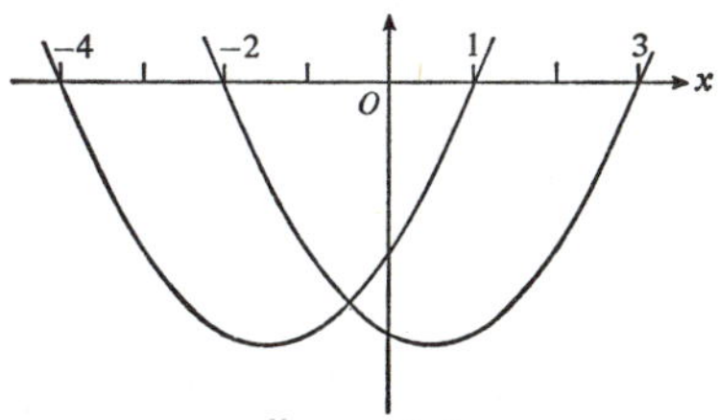

3.18 Inverting a Formula

If x is the required subject of a given formula for an expression y, then, provided that x appears only once, we can proceed as follows:

1. List in order the operations which must be carried out on x to produce the given expression.
2. Beside the list set the inverse of each operation.
3. Carry out these operations in reverse order on y.

Examples:

(1) $y = \sqrt{\left(\dfrac{3x^2-4}{5}\right)}$

x (↓)	y (↑)
square	$\sqrt{}$
$\times 3$	$\div 3$
-4	$+4$
$\div 5$	$\times 5$
$\sqrt{}$	square

$$\therefore x = \sqrt{\left(\frac{5y^2+4}{3}\right)}$$

(2) $Ps = \dfrac{W}{2g}(v^2 - u^2)$

↓	↑
u	Ps
square	$\sqrt{}$
$\times -1$	$\times -1 (= \div -1)$
$+v^2$	$-v^2$
$\times W$	$\div W$
$\div 2g$	$\times 2g$

$$u = \sqrt{\left\{-1\left(\frac{2gPs}{W} - v^2\right)\right\}}$$

$$\therefore u = \sqrt{\left\{v^2 - \frac{2gPs}{W}\right\}}$$

(3) $I = \dfrac{nE}{R+nr}$

↓	↑
r	I
$\times n$	$\div n$
$+R$	$-R$
invert	invert
$\times nE$	$\div nE$ *

The step marked * gives $\dfrac{I}{nE}$.

$$r = \frac{1}{n}\left(\frac{nE}{I} - R\right)$$

3.19 Multiplication of Triangular Matrices

We call a square matrix $A = (a_{pq})$ triangular if all its elements below the leading diagonal are zero, i.e. if $a_{pq} = 0$ when $p > q$.

I. *The product of two triangular matrices is triangular*

If $AB = C$, then we have (with the summation convention), $c_{pq} = a_{pr}b_{rq}$. Suppose $p > q$. When $r < p$, $a_{pr} = 0$; when $r \geqq p$, $b_{rq} = 0$. Hence $c_{pq} = 0$.

II. *The adjoint M of a triangular matrix A is triangular*

Apart from sign, m_{pq} is equal to the minor of a_{qp}, i.e. to the determinant of the matrix X got from A by substituting 1 for a_{qp}, and 0 for every other element of its row or column. If $p > q$, a_{qp} is above the leading diagonal, and X is triangular, and its determinant is equal to the product of the elements of its leading diagonal. Two of these are zero, namely those replacing a_{qq} and a_{pp}; hence $m_{pq} = 0$.

Corollary 1. *If A is triangular and non-singular, A^{-1} is triangular.*

Corollary 2. *If A is triangular and non-singular, and AB or BA is triangular, B is triangular.*

For $$B = A^{-1}(AB) = (BA)A^{-1}.$$

3.20 Illustrations for $x^0 = 1$

The average pupil is probably inclined to view with suspicion this curious result of his study of index laws in the algebra class. Teachers may like to know of a concrete way of illustrating it.

A fairly large sheet of uncreased paper, rectangular in shape, is folded repeatedly so that each fold divides the surface into two equal rectangles. The class is asked to observe that if the sheet is opened out again after any folding operation, it is found to be divided by the creases into a number of parts equal to some power of 2. Thus 1 fold gives 2^1 compartments, 2 folds give 2^2 compartments, and so on; n folds give 2^n compartments. It only remains for the class to realise that for the original unfolded sheet n was equal to zero, and that it contained 1 compartment, for it to be convinced that $2^0 = 1$.

In a similar way, if each folding operation consists of dividing the rectangular surface into 3 equal rectangles, it can be shown that on opening out the sheet after $1, 2, 3 \ldots n$ operations, the sheet is creased into $3^1, 3^2, 3^3, \ldots 3^n$ compartments; and again $n = 0$ corresponds to the original uncreased sheet having a single compartment. Hence $3^0 = 1$.

The extension, on similar lines, to the case x^n (where each operation consists of folding the surface into x equal areas) is a simple matter for the teacher. Here again $n = 0$ corresponds to the unfolded sheet having 1 compartment.

Another idea which might be useful is the following:

Is there any reason against defining x^5 (say) as 1 multiplied by x five times:

$$x^5 = 1 \times x \times x \times x \times x \times x?$$

This method of definition will enable the pupil easily to avoid incorrect statements such as '$x^2 = x$ multiplied by itself twice'; and the interpretation of x^0 as 1 seems to follow more naturally from this definition.

3.21 An Example on Mathematical Induction

In his *Mathematics and Plausible Reasoning* Professor G. Polya has remarked (Vol. I, p. 119) that a proof by mathematical induction may fail not only if one tries to prove too much but also if one tries to prove too little; and he illustrates the latter possibility by an example on the properties of a recursively defined function. The following example demonstrates the same point and has the added advantage of depending only on elementary school algebra.

If $a_n = \dfrac{1.3.5\ldots(2n-1)}{2.4.6\ldots(2n)}$ for n a positive integer, and if $a_k < \dfrac{1}{\sqrt{(3k)}}$ for some positive integer k, then

$$a_{k+1} < \frac{1}{\sqrt{(3k)}} \cdot \frac{2k+1}{2k+2} = \frac{1}{\sqrt{(3k+3)}}\sqrt{\frac{4k^2+4k+1}{4k^2+4k}},$$

from which one obviously cannot deduce that $a_{k+1} < \dfrac{1}{\sqrt{(3k+3)}}$.

On the other hand if $a_k \leqq \dfrac{1}{\sqrt{(3k+1)}}$ then

$$\begin{aligned} a_{k+1} &\leqq \frac{1}{\sqrt{(3k+1)}} \cdot \frac{2k+1}{2k+2} \\ &= \frac{1}{\sqrt{(3k+4)}} \cdot \sqrt{\frac{12k^3+28k^2+19k+4}{12k^3+28k^2+20k+4}} \\ &< \frac{1}{\sqrt{(3k+4)}}. \end{aligned}$$

It is thus possible to prove directly by induction that $a_n \leqq \dfrac{1}{\sqrt{(3n+1)}}$ but not that $a_n < \dfrac{1}{\sqrt{(3n)}}$.

It may also be noted that:

(i) the inductive proof of $a_n \geqq \dfrac{1}{\sqrt{(4n)}}$ presents no difficulty;

(ii) as $n \to \infty$, $a_n \sim \dfrac{1}{\sqrt{(\pi n)}}$ by Wallis's formula or Stirling's approximation,

(iii) the above upper bound for a_n can be used in conjunction with the comparison test to prove the convergence of the Maclaurin series for arc sin x at the endpoints of the range $(-1, 1)$ without appealing to Raabe's test.

The upper and lower bounds for a_n can also be used to establish the convergence for $-1 \leqq x < 1$ of the series $1 + \sum_{n=1}^{\infty} a_n{}^2 x^n$, the sum of which is a multiple of the complete elliptic integral of the first kind, viz. $\frac{2}{\pi}\int_0^{\pi/2} \frac{d\theta}{\sqrt{(1 - x\sin^2\theta)}}$.

A very simple example of a result too weak for induction is the inequality $(1+x)^n \geqq nx$, $x > 0$; for from $(1+x)^k \geqq kx$ follows $(1+x)^{k+1} \geqq (k+1)x - x(1-kx)$, and $1 - kx > 0$ only if $k < 1/x$. Of course the stronger result $(1+x)^n \geqq 1 + nx$, $x > 0$, may readily be proved by induction.

3.22 The Mixture as Before

To what familiar problem in elementary arithmetic does the identity $a(b-c)+b(c-a)+c(a-b) = 0$ supply the answer?

If Indian tea costs ap per kg and China tea bp per kg, what are the proportions of Indian and China tea in a blend costing cp per kg?

Since each kg of Indian tea costs $(c-a)$p *less* than the mixture, and each kg of China costs $(b-c)$p *more* than the mixture, therefore $(b-a)$ kg of the mixture must contain $(b-c)$ kg of Indian and $(c-a)$ kg of China tea, that is to say

$$a(b-c)+b(c-a) = c(b-a).$$

3.23 Crossing the Desert

A lorry when fully loaded can carry enough fuel to take it half-way across a desert. What is the minimum amount of fuel required to take it across the desert? It is assumed that any amount of fuel can be taken from the lorry at any point in the desert and this amount will remain undiminished till subsequently collected.

Solution, $7\frac{2021}{3002}$ full loads of fuel.

Proof. Consider in any solution the part of the desert between two successive storage points X and Y. Suppose the lorry travels from X to Y n times. Then it will travel from Y to X $(n-1)$ times and we shall say that the part of the desert between X and Y consumes fuel at $(2n-1)$ times the normal rate. Further if k is the total amount of fuel (in full load units) which during the whole journey starts out from X in the lorry in the direction X to Y, then $n \geqq k$.

Now divide the desert up by the points A, B, $C \ldots G$ as shown in the figure where the distance between the points are marked in units of half a desert width. P is the starting point and Q the end of the journey and

$$a = 1-\tfrac{1}{3}-\tfrac{1}{5}-\tfrac{1}{7}-\tfrac{1}{9}-\tfrac{1}{11}-\tfrac{1}{13} = \tfrac{2021}{3003} \times \tfrac{1}{15}$$

$$\langle\, a \,\rangle \langle \tfrac{1}{13} \rangle \langle \tfrac{1}{11} \rangle \langle \tfrac{1}{9} \rangle \langle \tfrac{1}{7} \rangle \langle \ldots \tfrac{1}{5} \ldots \rangle \langle \ldots \tfrac{1}{3} \ldots \rangle \langle \ldots\ldots 1 \ldots\ldots \rangle$$

$P \quad G \quad F \quad E \quad D \quad C \quad B \quad A \quad Q$

Let $t(S)$ be the algebraic total amount of fuel (in full load units) which during the whole journey passes the point (a general point) in the lorry measured positively in the direction S to Q. Let N be the normal rate of consumption of fuel. Then any part of the desert nearer to P than S must consume fuel at a rate $\geqq [2t(S)-1]N$.

Certainly any part of AQ must consume at a rate $\geqq N$ and hence $t(A) \geqq 1$. Consequently $t(S) > 1$ if S lies nearer to P than A, and hence any part of BA must consume at a rate $\geqq 3N$. Therefore

$$t(B) \geqq \tfrac{1}{3} \times 3 + t(A) \geqq 2.$$

Consequently $t(S) > 2$ if S lies nearer to P than B, and hence any part of CB must consume at a rate $\geqq 5N$. Therefore

$$t(C) \geqq \tfrac{1}{5} \times 5 + t(B) \geqq 3.$$

Continuing in this way we obtain $t(G) \geqq 7$. Consequently $t(S) > 7$ if S lies between P and Q, and hence any part of PQ must consume at a rate $\geqq 15N$. Therefore

$$t(P) \geqq a \times 15 + t(G) \geqq 7\tfrac{2021}{3003}.$$

It is easy to see that this minimum can be attained. The simplest method is to use A, B, C, D, E, F and G as the only storage points and start out from P 7 times fully loaded, leaving $(1-2a)$ fuel units at G each time and returning to P. Start out from P the eighth time with the remaining fuel, $15a$ units, and thus arrive at G with a complete total of 7 units. We can now start out from G 7 times fully loaded, leaving $\tfrac{11}{13}$ fuel units at F on each of the first 6 occasions and returning to F. We thus arrive at F for the seventh time with a complete total of 6 units. Continuing in this way from F to E and E to D etc., we eventually arrive at A for the second time with $\tfrac{2}{3}$ units on the lorry and $\tfrac{1}{3}$ units already unloaded at A; just enough to fully load the lorry and take it to Q.

REFERENCES

3.1. *Solution of Quadratic Equations by Geometrical Constructions*

H. G. Middleton 1908, Vol. 30, No. 290, p. 151
R. L. Goodstein C 59, Vol. 44, No. 350, p. 289
J. W. Hesselgreaves C 1, Vol. 42, No. 339, p. 57

3.2. *A Graphical Representation of Quadratic Equations*

H. Gebert 2549, Vol. 39, No. 329, p. 232

3:3. *A Nomogram for the Solution of Quadratic Equations*

C. V. Gregg 2529, Vol. 39, No. 328, p. 153

3.4. *An Interesting Equation*

T. Nakazawa 3038, Vol. 46, No. 358, p. 315
H. M. Finucan 3090, Vol. 48, No. 363, p. 93

3.5. *The Condition for a Cubic Equation to have Equal Roots*

D. B. Eperson 3336, Vol. 56, No. 397, p. 232

3.6. *The Real Roots of a Cubic Equation*

A. R. Pargeter C114, Vol. 48, No. 364, p. 211

3.7. *Complex Roots of a Cubic Equation from a Graph*

F. H. Francis C214, Vol. 54, No. 388, p. 145

3.8. *Graphs of Cubic Functions*

A. Hurrell 2635, Vol. 40, No. 334, p. 274
T. Nakazawa C 39, Vol. 43, No. 345, p. 203
D. G. Bousfield C 56, Vol. 44, No. 349, p. 197
E. J. F. Primrose C 57, Vol. 44, No. 349, p. 197

3.9. *The Remainder Theorem*

A. G. Sillitto C 23, Vol. 42, No. 341, p. 211
H. G. Woyda C 50, Vol. 44, No. 347, p. 61

3.10. *The Inequality between the Arithmetic and Geometric Means*

B. E. Lawrence 1155, Vol. 19, No. 234, p. 221
V. Popovic 2237, Vol. 35, No. 314, p. 246

3.11. *Applications of the Inequality of the Means*

J. Sinnadura 2979, Vol. 45, No. 354, p. 330

3.12. *The Binomial Theorem for any Index*

A. J. Carr 2359, Vol. 37, No. 322, p. 266

3.13. *Pascal's Triangle*

W. A. Broomhead C248, Vol. 55, No. 393, p. 317
G. A. Garreau 1519, Vol. 25, No. 264, p. 118
B. Cohen 2381, Vol. 37, No. 322, p. 289
F. W. Kellaway 1762, Vol. 28, No. 282, p. 190
A. R. Pargeter C129, Vol. 49, No. 367, p. 76
J. L. Atkin 2901, Vol. 44, No. 349, p. 199

3.14. *Sums of Powers of the Natural Numbers. Proofs by Dissection Methods*

A. P. Rollett 1262, Vol. 21, No. 247, p. 412
F. C. Boon 1263, Vol. 21, No. 247, p. 414
E. H. Lockwood C300, Vol. 58, No. 403, p. 50
N. R. C. Dockeray 1330, Vol. 22, No. 252, p. 488

3.15. *Sums of Powers of the Natural Numbers. Proofs by Algebraic Methods*

T. W. Chaundy 1541, Vol. 25, No. 266, p. 240
R. F. Wheeler 2687, Vol. 41, No. 336, p. 122
G. T. Whitrow C 16, Vol. 42, No. 340, p. 116

3.16. *A Single-Scale Nomogram*
R. K. Guy 2315, Vol. 37, No. 319, p. 39
3.17. *The Sign of a Rational Function*
J. W. Gubby C291, Vol. 57, No. 401, p. 203
3.18. *Inverting a Formula*
B. M. Read C143, Vol. 50, No. 373, p. 299
3.19. *Multiplication of Triangular Matrices*
M. F. Egan 2586, Vol. 40, No. 331, p. 44
3.20. *Illustration for* $x^0 = 1$
W. G. Guthrie 1830, Vol. 29, No. 285, p. 121
B. Evans 2074, Vol. 33, No. 306, p. 281
3.21. *An Example on Mathematical Induction*
A. V. Boyd C 73, Vol. 45, No. 353, p. 248
3.22. *The Mixture as Before*
R. L. Goodstein 2039, Vol. 33, No. 303, p. 32
3.23. *Crossing the Desert*
G. G. Alway 2707, Vol. 41, No. 337, p. 209

4 Calculus and Analysis

4.1 An Area Property of the Curve $y = x^n$

The area bounded by an arc of the curve $y = x^n$ the abscissae through its end points and the y-axis is n times the area bounded by the arc, the ordinates through its end points and the x-axis.

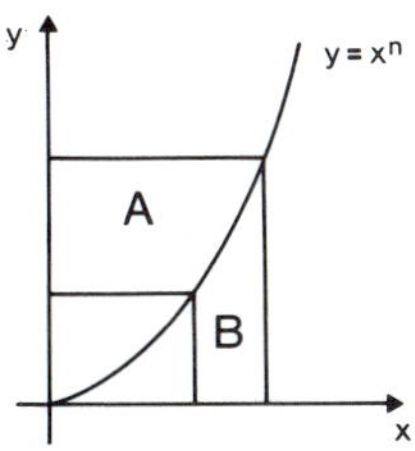

A =nB

For the second area is

$$\int_a^C y\,dx = \frac{1}{n+1}(C^{n+1} - a^{n+1})$$

and the first is

$$\int_{a^n}^{C^n} x\,dy = \frac{n}{n+1}(C^{n+1} - a^{n+1}).$$

4.2 Extending the Trapezial Rule for Area

The scope of the formula $\frac{1}{2}(a+b)h$ can be extended to include areas other than the trapezium.

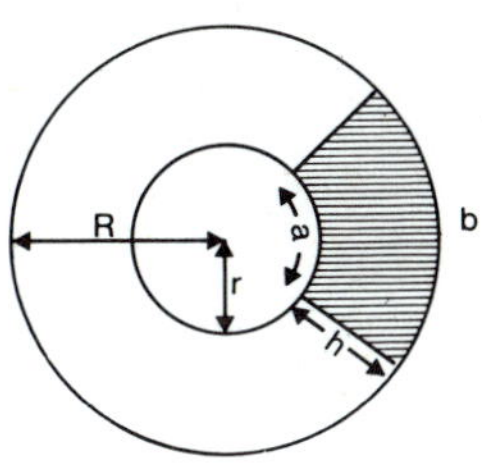

SHADED AREA
= ½ (a+b) h

The area between arcs of lengths a and b ($b > a$) of concentric circles bounded by normals may be derived by dividing both arcs into n equal parts, and joining the $(n-1)$ corresponding points on each arc to form approximately n equal trapezia, of parallel sides a/n and b/n and distance apart $h = R-r$, so that total area becomes $n \,.\, \frac{1}{2}(a/n+b/n)h = \frac{1}{2}(a+b)h$. Hence also for the curved surface of the frustum of a right circular cone, with a and b circumferences of the ends and h the slant height.

With $a = 0$ and $h = R$, area of sector is $\frac{1}{2}bR$.

With $a = 2\pi r$, $b = 2\pi R$ and $h = R-r$, area between concentric circles is $\frac{1}{2}(2\pi r+2\pi R)(R-r) = \pi(r+R)(R-r)$—in the form suitable for computation.

With $a = 0$, $b = 2\pi R$, $h = R$, area of circle is $\frac{1}{2}(0+2\pi R)R = \pi R^2$.

4.3 Functions, Differences and the Fundamental Theorem of the Calculus

This note describes the place of tabulated functions in an elementary mathematics course. Much of it will be well known, but the approach to integration does not appear to be. It provides a method of linking antidifferentiation with the integral as the limit of a sum.

The study of tables of functions is important for the sake of numerical work. It can also be used to illuminate an introduction to the calculus, providing some students with a more powerful stimulus than the graphical work that goes with it.

The form of elementary table most common in schools has two-dimensional tabulation, and mean proportional parts for each row of the table. This is convenient and practical for use, but is not the best form for introducing the study of tables. A better form of tabulation is a listing in a single column, with the first differences given in units of the last decimal place of tabulation. Chambers' four-figure tables by Comrie include some functions tabulated in this way, and are ideal for an elementary course.

In the initial study of tables, a wider interval of tabulation than would be used in published tables is often useful. As an example, here is a table of $\sin x$ for $x = 0{\cdot}8(0{\cdot}1)1{\cdot}2$, 4D.

x	$\sin x$		
0·8	0·7174		
		659	
0·9	0·7833		−77
		582	
1·0	0·8415		−85
		497	
1·1	0·8912		−89
		308	
1·2	0·9320		

The second differences of the function have also been given.

The relation of the table to the graph of the function is illuminating, the increase of the function giving positive differences, and the decreasing differences showing that the graph will be concave downwards.

The use of differences in linear interpolation, with the idea of replacing the curve by a chord, brings in ratio and proportion, and similar triangles. This can lead to an understanding of tables with mean proportional parts, and a critical discussion of their use.

Differentiation

The relation $\delta y \simeq y'\delta x$ is illustrated by the table (in the particular table given, the differences should be compared with values of $\cos x$). Second and higher order differences can be formed and their relation to higher derivatives explored. In the table given, the second differences clearly approximate $-(0{\cdot}1)^2 \sin x$. A further useful step is to give a table of values of displacement along a line given at equal intervals in time, and relate the first differences to average velocity.

Integration

In what follows it is assumed that the definition of $\int_a^b y\,dx$ as the limit of a sum has been given and used, but not yet linked to antidifferentiation.

Moving one column to the right in a difference table corresponds approximately to differentiation. It follows that moving to the *left* should correspond to antidifferentiation. Since building up the function values from the differences involves summation, the link between antidifferentiation and summation is suggested. If the differences only are given, then a function value must be chosen arbitrarily before the function can be reconstructed—this links with a graphical discussion, and with the analytical problem.

The fundamental theorem

To make the connection more explicit we note the exact relation:

$$\sum_{x=a}^{b} \delta y = y(b) - y(a) = [y(x)]_a^b. \tag{1}$$

This will be obvious from the meaning of δy—it is useful as a check that there are no mistakes in forming a table of differences.

Noticing that $\delta y \simeq y'\delta x$, (1) immediately suggests

$$\int_a^b y'\,dx = y(b) - y(a).$$

To make the argument more satisfying (where complete rigour will defer to understanding) we have:

$$y' = \operatorname*{Lt}_{\delta x \to 0} \frac{\delta y}{\delta x} \Leftrightarrow \frac{\delta y}{\delta x} = y' + \varepsilon$$

where $\varepsilon \to 0$ as $\delta x \to 0$, i.e.

$$\delta y = y'\delta x + \varepsilon\delta x;$$

thus

$$y(b) - y(a) = \sum_{x=a}^{b} \delta y$$

$$= \sum_{x=a}^{b} y'\delta x + \delta x \sum_{x=a}^{b} \varepsilon. \tag{2}$$

(δx remains constant through the whole table).

If E is the greatest of the $|\varepsilon|$ for any given interval of tabulation, then $\left|\delta x \sum_{x=a}^{b} \varepsilon\right| < \delta x \,.\, nE = (b-a)E$, where $E \to 0$ as $\delta x \to 0$. n is the number of differences between $y(a)$ and $y(b)$.

Further

$$\operatorname*{Lt}_{\delta x \to 0} \sum_{x=a}^{b} y'\delta x = \int_a^b y'\,dx \qquad \text{(by definition).}$$

Proceeding to the limit as $\delta x \to 0$ in (2) thus gives

$$y(b) - y(a) = \int_a^b y'\,dx. \tag{3}$$

Since this is the form of the fundamental theorem of the calculus that is actually used to evaluate integrals, it seems the best form to arrive at in a first course—better than

$$\frac{d}{dx}\left\{\int_a^x f(t)\,dt\right\} = f(x),$$

which is the form more commonly derived (e.g. by differentiating the area under a curve).

The final step in this exploration might be to find the area of a triangle, already shown to be $\int_0^1 x\,dx$, and already evaluated both by geometry and by summation of series, as follows:

$$\int_0^1 x\,dx = \int_0^1 \frac{d}{dx}\left(\frac{x^2}{2}\right)dx = \left[\frac{x^2}{2}\right]_0^1 = \tfrac{1}{2}.$$

4.4 How Arbitrary can a Constant of Integration be?

A possible introduction to 'what is an indefinite integral?' Integration by parts gives

$$\begin{aligned}\int \tan x\,dx &= \int \sec x \,.\, \sin x\,dx \\ &= \sec x \,.\, (-\cos x) - \int \sec x \tan x \,.\, (-\cos x)\,dx + c,\end{aligned}$$

where c is an arbitrary constant. Hence

$$\int \tan x\,dx = -1 + \int \tan x\,dx + c$$

and thus $c = 1$, contrary to the fact that it is arbitrary.

4.5 Integration of Inverse Functions

For the integral of the inverse $f^{-1}(x)$ of a function $f(x)$, whose integral is known, one method of proceeding is first to substitute $x = f(y)$ and then to integrate by parts as follows:

$$\int_a^b f^{-1}(x)\,dx = \int_{f^{-1}(a)}^{f^{-1}(b)} f^{-1}[f(y)]\,df(y) = \int_{f^{-1}(a)}^{f^{-1}(b)} y\,df(y)$$

$$= [yf(y)]_{f^{-1}(a)}^{f^{-1}(b)} - \int_{f^{-1}(a)}^{f^{-1}(b)} f(y)\,dy$$

$$= bf^{-1}(b) - af^{-1}(a) - \int_{f^{-1}(a)}^{f^{-1}(b)} f(y)\,dy. \qquad (1)$$

Another method of proceeding is to integrate by parts first and then to substitute in the integral obtained.

If one thinks of this problem geometrically, the derivation of equation (1) becomes even easier. Consider the graph of the function

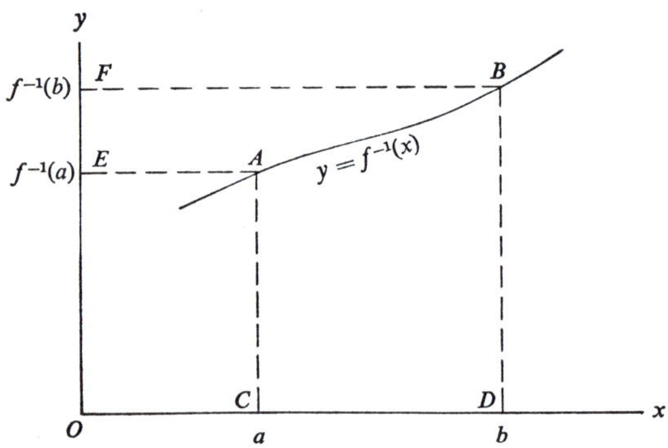

$y = f^{-1}(x)$ between $x = a$ and $x = b$ as shown in the figure. Then

$$\int_a^b f^{-1}(x)\,dx = \text{area}\,(ABDC)$$

$$= \text{area}\,(OFBD) - \text{area}\,(OEAC) - \text{area}\,(EFBA)$$

$$= bf^{-1}(b) - af^{-1}(a) - \int_{f^{-1}(a)}^{f^{-1}(b)} f(y)\,dy,$$

which agrees with equation (1).

This method also applies to indefinite integrals, giving

$$\int^{x} f^{-1}(t)\,dt = xf^{-1}(x) - \int^{f^{-1}(x)} f(t)\,dt.$$

4.6 Some Integration Techniques

I A format for repeated integration by parts.

Integrals of the form $\int x^n f(x)\,dx$ are often evaluated by repeated integration by parts. When n is a positive integer, the tabular layout illustrated in the following example has been found useful.

Example. $I = \displaystyle\int x^5 \sin x\,dx.$

SIGN	*DERIVATIVES*	*INTEGRALS*
		$\sin x$
$+$	x^5	$-\cos x$
$-$	$5x^4$	$-\sin x$
$+$	$20x^3$	$\cos x$
$-$	$60x^2$	$\sin x$
$+$	$120x$	$-\cos x$
$-$	120	$-\sin x$

Then

$$I = -x^5\cos x + 5x^4\sin x + 20x^3\cos x - 60x^2\sin x - 120x\cos x + 120\sin x.$$

II The second technique fits into the normal sequence for the integration of fractions involving polynomials, and requires the learning of two fairly simple 'formulae' which replace innumerable so-called standard forms.

To integrate with respect to x the function $1/f(x)$ where $f(x)$ is a quadratic ax^2+bx+c, the symbol ∇ (NABLA) is used to represent $\sqrt{|b^2-4ac|}$ (the square root of the modulus of the discriminant). The special case when $b^2-4ac=0$ occurs when $f(x)$ is a perfect square and the integral is then a standard form. Otherwise:

(i) If $b^2-4ac>0$, $\displaystyle\int \frac{1}{f(x)}\,dx = \frac{1}{\nabla}\log_e \frac{f'(x)-\nabla}{f'(x)+\nabla}$;

(ii) if $b^2-4ac<0$, $\displaystyle\int \frac{1}{f(x)}\,dx = \frac{2}{\nabla}\tan^{-1}\frac{f'(x)}{\nabla}$.

There is a pattern about these two formulae that makes them fairly simple to memorise although one can use the celebrated tune 'Tea for Two' as a mnemonic for the coefficient numerators (t for $\tan^{-1}$... or arctan if this is popular!).

Views may differ as to the advisability of such a 'mechanical' process but, for a step which is (for instance, in engineering problems) only incidental to a more important problem, an efficient and quick completion by inspection is a great boon.

III The integration of $\sec x$ and $\sec^3 x$.

$$\begin{aligned}
&\text{If} & \sec x &= \cosh v,\\
&\text{then} & \sec^2 x - 1 &= \cosh^2 v - 1\\
&\text{and} & \tan x &= \sinh v,\\
&\text{so that, since} & \sec x \tan x\, dx &= \sinh v\, dv,\\
&\text{we have} & \sec x\, dx &= dv\\
&\text{and} & \textstyle\int \sec x\, dx &= \textstyle\int dv = v\\
&& &= \operatorname{arg\,cosh}(\sec x)\\
&& &= \log(\sec x + \tan x).
\end{aligned}$$

$$\begin{aligned}
&\text{Also,} & \textstyle\int \sec^3 x\, dx &= \textstyle\int \cosh^2 v\, dv\\
&& &= \tfrac{1}{2}\textstyle\int (1+\cosh 2v)\, dv\\
&& &= \tfrac{1}{2}(v + \tfrac{1}{2}\sinh 2v)\\
&& &= \tfrac{1}{2}\{\operatorname{arg\,cosh}(\sec x) + \sec x \tan x\}.
\end{aligned}$$

A further method of integrating $\sec^3 x$ found by an ex-secondary modern pupil:

$$\begin{aligned}
2\int \sec^3 x\, dx &= \int (\sec x + \sec x \tan^2 x + \sec^3 x)\, dx\\
&= \int \sec x\, dx + \int d(\sec x \tan x)\\
&= \log(\sec x + \tan x) + \sec x \tan x + c.
\end{aligned}$$

IV Finally, a method of integrating the product of two special functions, without using integration by parts.

If $f(x)$, $g(x)$ be two functions satisfying the relation

$$\frac{f''(x)}{f(x)} - \frac{g''(x)}{g(x)} = k,$$

then it is easily verified that

$$\int f(x)g(x)\,dx = \{f'(x)g(x)-f(x)g'(x)\}/k.$$

This formula is specially serviceable for the following integrals:

$$\int e^{ax} \begin{matrix}\sin\\ \cos\end{matrix}\, bx\,dx; \quad \int e^{ax} \begin{matrix}\sinh\\ \cosh\end{matrix}\, bx\,dx;$$

$$\int \begin{matrix}\sin\\ \cos\end{matrix}\, mx \begin{matrix}\cosh\\ \sinh\end{matrix}\, nx\,dx; \quad \int \begin{matrix}\sin\\ \cos\end{matrix}\, mx \begin{matrix}\sin\\ \cos\end{matrix}\, nx\,dx;$$

$$\int \begin{matrix}\sinh\\ \cosh\end{matrix}\, mx \begin{matrix}\sinh\\ \cosh\end{matrix}\, nx\,dx.$$

Example. $\int e^{ax} \cosh bx\,dx.$

Put $f(x) = e^{ax}$, $g(x) = \cosh bx$, then

$$f''(x)/f(x) = a^2, \quad g''(x)/g(x) = b^2,$$

whence $$k = a^2-b^2.$$

Hence

$$\int e^{ax}\cosh bx\,dx = \frac{e^{ax}}{a^2-b^2}(a\cosh bx - b\sinh bx).$$

4.7 An Old Problem: a Donkey Tethered to the Edge of a Circular Field

The problem of the donkey tethered to the edge of a circular field appears to be of great interest to the student doing Newton's method for solving equations. The following method for obtaining the area involved may be of interest as being more in accordance with the work on calculus that usually precedes this stage of teaching.

The circular field of radius r has centre O. (See fig. opposite). The donkey is tethered at R by a rope of length PR. Then in the figure $2r\cos\theta = x$, and if we denote the angle QRP by α, the area common to the two circles is

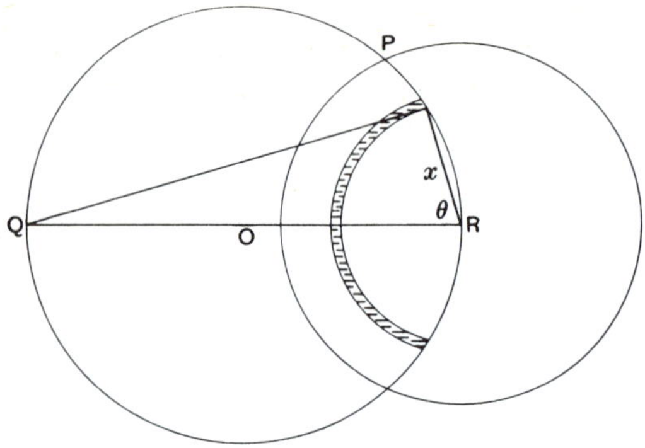

$$\int 2x\theta\, dx = -4r^2 \int_{\pi/2}^{\alpha} \theta \,.\, \sin 2\theta \,.\, d\theta$$

$$= 2r^2\{\alpha \cos 2\alpha - \tfrac{1}{2}\sin 2\alpha + \tfrac{1}{2}\pi\}.$$

If this area is to be the fraction k of the whole area πr^2 of the field, we obtain the usual equation

$$2\alpha \cos 2\alpha - \sin 2\alpha + \pi(1-k) = 0.$$

4.8 Investigating Stationary Points

When investigating the stationary points of a given function, it is often advisable to consider the sign of dy/dx near the point rather than evaluating d^2y/dx^2.

For example,

$$y = \frac{(x-1)^2(x+2)}{(x-2)^2},$$

and

$$\frac{dy}{dx} = \frac{(x-1)(x^2-5x-2)}{(x-2)^3},$$

$$= \frac{(x-1)(x-\alpha)(x-\beta)}{(x-2)^3},$$

where $\alpha = \frac{1}{2}(5+\sqrt{33})$, $\beta = \frac{1}{2}(5-\sqrt{33})$.

Note that $\beta < 1 < 2 < \alpha$.

Then if $x = \beta-$(i.e. x is just less than β) $dy/dx > 0$, but if $x = \beta+$ then $dy/dx < 0$. Thus $x = \beta$ gives a local maximum. Similarly $x = 1$ and $x = \alpha$ are local minima.

This approach has the added advantage that it gives the student practice in simple inequalities, which are of ever-increasing importance.

4.9 The Gradient of the Sine Curve

The following proof of the result $\frac{d}{dx}(\sin x) = \cos x$ was suggested and worked out by a sixth former. It does not require a knowledge of the factor formulae in trigonometry or of the limit $(\sin \theta)/\theta$ as $\theta \to 0$.

If a square piece of paper of side 2π units is wrapped round a cylinder of unit radius, a diagonal of the square is transformed into one circuit of a helix which cuts all generators at an angle of $\frac{1}{4}\pi$ radians. In Fig. 1, the coincident edges of the square lie along the line AB, which is taken for convenience to be in the horizontal plane through the axis of the cylinder. The broken line in the figure is the projection of this helix on the vertical plane through the axis of the cylinder; it can be shown that this has the form of a sine curve:

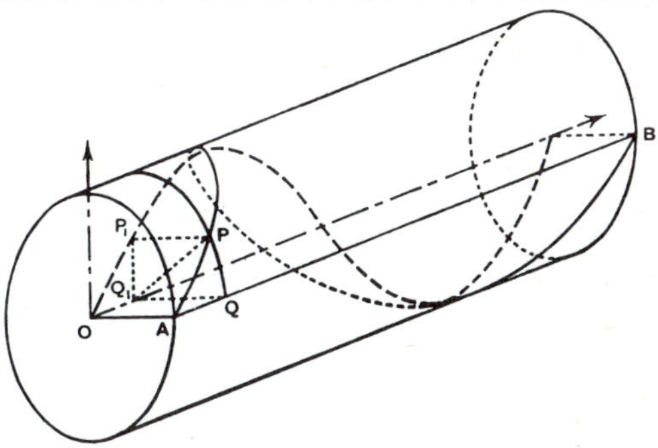

FIG. 1

The proof of this is as follows. Let P be a general point of the helix, and PQ a line which was parallel to an edge of the square before it was wrapped round the cylinder. P_1, Q_1 are the projections of P, Q on the vertical plane. Then if $AQ = x$, we have $PQ = x$ and angle $PQ_1Q = x$ radians, since the cylinder has unit radius. Therefore angle $Q_1PP_1 = x$ radians, and $P_1Q_1 = \sin x$. Since $OQ_1 = x$, P_1 lies on a sine curve.

In Fig. 2, P_1T_1 is the tangent to the sine curve and PT the tangent to the helix, so that T_1 is the projection of T. PN is tangent to the arc

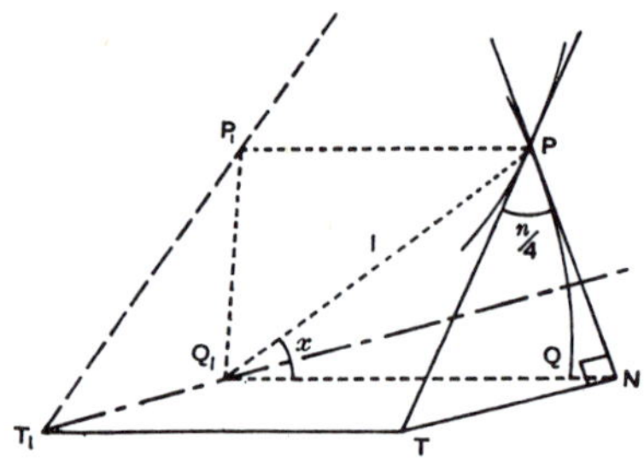

FIG. 2

PQ, so that the plane PNT touches the generator through P. Therefore TN is parallel to the axis of the cylinder, and also $TN = T_1Q_1$. Since the helix has an angle of inclination of $\frac{1}{4}\pi$ radians, $\angle TPN = \frac{1}{4}\pi$, so that $NP = TN$. It follows that $NP = T_1Q_1$.

Now, from triangle PQ_1N,

$$NP = \tan x,$$

so that $T_1Q_1 = \tan x$. But it has already been proved that $Q_1P_1 = \sin x$. Therefore the gradient of the sine curve is

$$\frac{Q_1P_1}{T_1Q_1} = \frac{\sin x}{\tan x} = \cos x.$$

The device for differentiating the sine-function described here is equivalent to the well-known device of considering motion round a circle. If P moves round the unit circle with centre at the origin at a rate of 2π revolutions per unit time, starting at $(1, 0)$ when $t = 0$, then its coordinates at time t will be $(\cos t, \sin t)$. Hence the vertical component of its velocity will be $D_t \sin t$. But it is then travelling at an angle t to the y-axis and so the vertical component of the velocity is also $\cos t$.

It is in quite a strict sense that the two devices are equivalent, because the graph of the motion of P, with t-axis at right-angles to the x- and y-axes, is precisely the helix considered earlier.

4.10 Differentiation of Logarithms

We define a logarithm from the relationship

$$\log(p \,.\, q) = \log p + \log q \tag{1}$$

This equation leads at once to the fact that

$$\log 1 = 0.$$

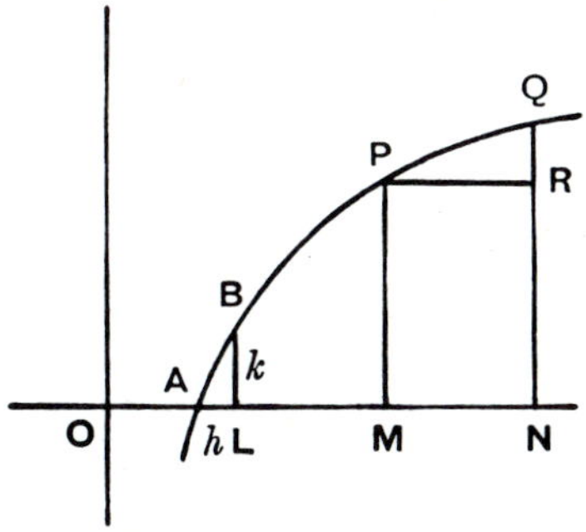

Thus the graph of the function

$$y = \log x$$

crosses the axis of x at the point $A(1, 0)$.

Let B, P, Q be three points on this graph, and let L, M, N be the feet of the perpendiculars on the x-axis.

Also let
$$OL = 1+h,$$
$$OM = x,$$
$$ON = x(1+h).$$

Let
$$\log(1+h) = LB = k,$$
$$\log x = MP = y,$$
$$\log[x(1+h)] = NQ = NR+RQ = y+RQ.$$

Then, by virtue of (1),

$$\log[x(1+h)] = \log x + \log(1+h) = y + LB = y+k$$

$$\therefore \qquad RQ = k.$$

Also
$$ON = x(1+h) = x+xh,$$

so that
$$PR = MN = xh.$$

Now we have:

(i) gradient of chord $AB = \frac{k}{h}$, and thus gradient of tangent at $A = \lim_{h \to 0}\left(\frac{k}{h}\right)$;

(ii) gradient of chord $PQ = \frac{k}{xh}$, and thus gradient of tangent at $P = \lim_{h \to 0}\left(\frac{k}{xh}\right) = \frac{1}{x}\lim_{h \to 0}\left(\frac{k}{h}\right)$.

Now choose the base of our logarithms for $\lim\left(\frac{k}{h}\right)$ to be equal to 1; then we have $\frac{dy}{dx} = \frac{1}{x}$.

4.11 Another Approach to the Logarithmic and Exponential Functions

Let $F(x)$ be defined by the equation

$$\frac{d}{dx}[F(x)+c] = \frac{1}{x}, \tag{1}$$

where c is some constant; and consider the function inverse to $F(x)$, namely $y = E(x)$, which implies $x = F(y)$.

$$\frac{dx}{dy} = \frac{1}{y}, \quad \text{and so} \quad \frac{dy}{dx} = y. \tag{2}$$

We seek, together, the two functions, mutually inverse, which satisfy the equations (1) and (2).

Now

$$\frac{d}{dx}[F(x^n)] = \frac{nx^{n-1}}{x^n} = \frac{n}{x} = n\frac{d}{dx}[F(x)].$$

Thus $\qquad F(x^n) - nF(x) = \text{a constant, independent of } x,$

$$= F(1) - nF(1), \quad \text{putting } x = 1.$$

We now choose the constant c so that $F(1) = 0$ (and so $E(0) = 1$).

Then $\qquad F(x^n) = nF(x).$

Denoting $F(x)$ by z, so that $x = E(z)$, we have

$$F\{E(z)\}^n = nz,$$

so that

$$\{E(z)\}^n = E(nz).$$

Putting $z = 1$, this yields

$$\{E(1)\}^n = E(n).$$

Writing e for the number $E(1)$, we have

$$E(n) = e^n, \quad \text{and} \quad E(x) = e^x,$$

and so the inverse function, $F(x)$, is $\log_e x$.

4.12 Continuous Interest and the Exponential Function

On a growing amount £P, a function of t years, the interest from time $t = 0$ to t at the rate of simple interest £r per £ per annum is

$$\int_0^t Pr\,dt. \tag{1}$$

To find the total interest on £1 for t years at £r per pound per annum, the interest being added continuously. The simple interest on £1 for time $t = \int_0^t 1 \times r\,dt = rt$. This amount rt has been growing from 0 to t and the interest on it is

$$\int_0^t rt \times r\,dt = \frac{r^2t^2}{2},$$

and the interest in $r^2t^2/2$ growing from 0 to t is

$$\int_0^t \frac{r^2t^2}{2} \times r\,dt = \frac{r^3t^3}{3!}$$

and so on. Hence adding the original £1 the amount A at time t is

$$A = 1 + rt + \frac{r^2t^2}{2!} + \frac{r^3t^3}{3!} + \dots \tag{2}$$

The relation to e^{rt} is found in the usual way from

$$dA = Ar\,dt$$
$$\log A = rt + C,$$

so that if

$$A = 1 \text{ at } t = 0$$
$$\log A = rt$$
$$A = e^{rt}.$$

4.13 Limit of $\sin\theta/\theta$ as $\theta \to 0$

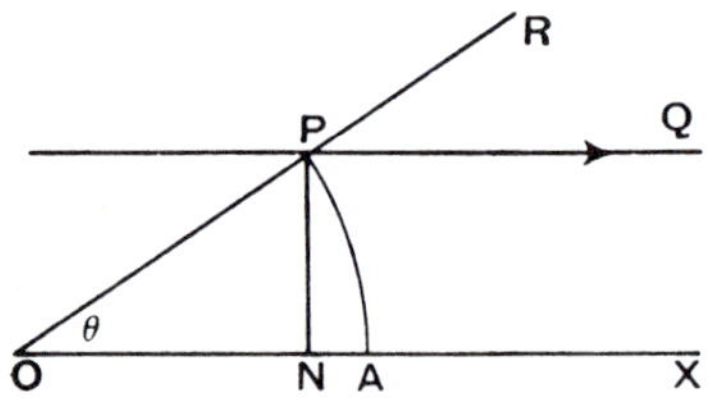

Let θ be the angle XOR. With centre O describe a circular arc AP, cutting OX in A and OR in P. Let PN be the perpendicular from P to OX. Then

$$\frac{\sin\theta}{\theta} = \frac{NP}{OP}\Big/\frac{\text{arc } AP}{OP} = \frac{NP}{\text{arc } AP}.$$

Now draw PQ parallel to OX and let P move 'to infinity' along PQ (so that NP remains fixed in length); then, plainly, the limit of the length of the arc AP is the length of NP; so that $NP/\text{arc } AP \to 1$. Also, $\theta \to 0$, and hence

$$(\sin\theta)/\theta \to 1 \quad \text{when} \quad \theta \to 0.$$

4.14 The Convergence of x^n

It is often the practice in textbooks to make quite heavy weather of the proof that $x^n \to 0$ in the interval $-1 < x < 1$, and the crossing for nx^n is even rougher. The usual argument writes $1/|x| = 1+\delta$, $\delta > 0$, and then appeals to the inequality $(1+\delta)^n \geqq 1+n\delta$ to obtain $|x|^n < 1/n\delta$; for nx^n we use the next term in the binomial to give $n|x|^n < 2/(n-1)\delta$. In fact, we do not need to look beyond the geometric series itself to find all that we require. For, if $0 < x < 1$, then

$$nx^n < x + x^2 + \ldots + x^n = (x - x^{n+1})/(1-x) < x/(1-x). \qquad (1)$$

Whence, if $0 < x < X$, taking x/X for x in (1), we have

$$nx^n < \{x/(X-x)\}X^n. \qquad (2)$$

From (1) it follows that $x^n \to 0$, and thence from (2) that $nx^n \to 0$. Multiplying (2) by n^k, we see that $n^{k+1}x^n \to 0$ if $n^k x^n \to 0$ (for $0 < x < 1$) which completes an inductive proof that $n^r x^n \to 0$ for any r.

4.15 The Harmonic Series

The following method of proving that $\sum \frac{1}{n}$ diverges is much shorter than that given in most textbooks:

We first note that if S_n is the nth partial sum of a series which converges to S then

$$\lim (S_{2n} - S_n) = S - S = 0 \qquad (1)$$

But for the series $\sum \frac{1}{n}$ we have

$$S_{2n} - S_n = \frac{1}{n+1} + \frac{1}{n+2} + \ldots + \frac{1}{2n} > n \cdot \frac{1}{2n}$$

i.e.

$$S_{2n} - S_n > \frac{1}{2}$$

so that (1) cannot hold.

However, if the denominators of the terms of the series are read as numbers in scale eleven (or any higher scale) the series is convergent, though rather slowly. The sum lies between 2·828 and 26·29.

4.16 A Topological Approach to Continuity

Students are always complaining that the definitions of limit and continuity demand a special capacity for intellectual contortionism. G. H. Hardy asked us to imagine ourselves confronted by an opponent who was determined to outwit us by choosing impossibly small values of 'epsilon'. Nowadays we are encouraged to think in more general terms of functions under which the inverse images of open sets are open sets—which to the beginner seems a 'back to front' way of looking at it.

Perhaps the fashion for drawing Venn diagrams and putting symbols in braces may help the present-day student of analysis. He knows that the proposition

all pigs are animals,

expressed rather more formally

$$x \text{ is a pig} \Rightarrow x \text{ is an animal}$$

can be illustrated by

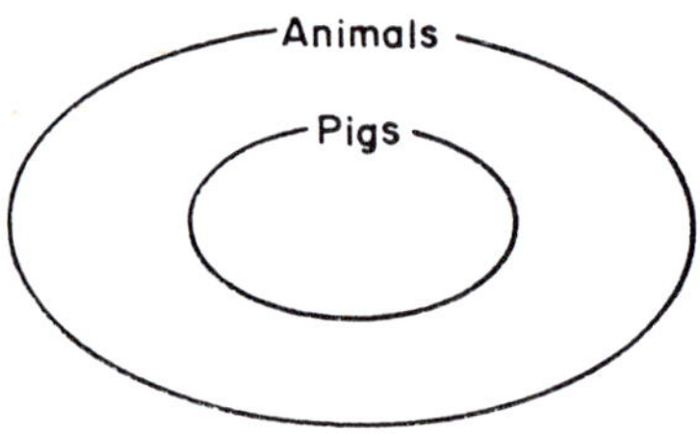

That is,

$$\{\text{pigs}\} \subset \{\text{animals}\},$$

or

$$\{x : x \text{ is a pig}\} \subset \{x : x \text{ is an animal}\}.$$

Let us apply this to continuity. For sixth formers and first year undergraduates it is fair enough to assume that we have metric spaces, so that we can speak of a point as being 'close' to another. Then the common sense meaning of continuity is that

$$x \text{ is close to } a \Rightarrow f(x) \text{ is close to } f(a);$$

and following our agricultural example above we may express this in the form

$$\{x : x \text{ is close to } a\} \subset \{x : f(x) \text{ is close to } f(a)\}.$$

We are now home. The set on the left (see opposite) is a neighbourhood S of a in the domain. Notice that the set on the right is also a subset of the domain: the set of points of the domain whose images under f are close to the image of a. We call this the 'inverse image' of a neighbourhood Σ of $f(a)$ in the codomain, and denote it by $f^{-1}(\Sigma)$.

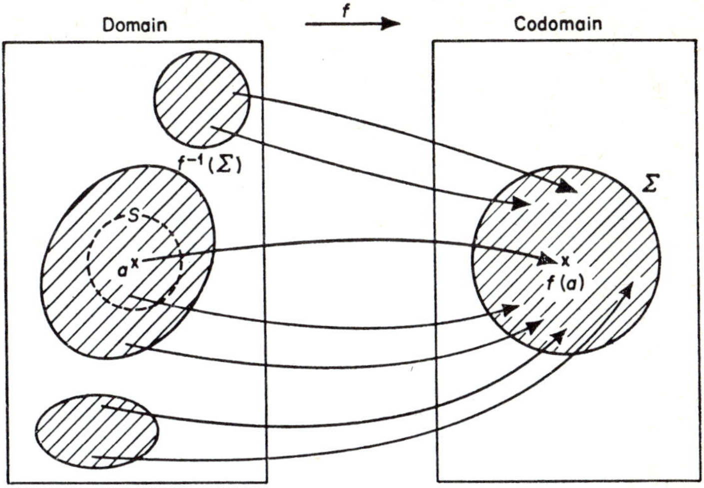

For continuity we require, therefore,

$$S \subset f^{-1}(\Sigma).$$

That is, there must exist a neighbourhood S of a which is contained in $f^{-1}(\Sigma)$, the inverse image of the neighbourhood Σ of $f(a)$.

We can, of course, get back immediately to our old friends ε and δ by taking domain and codomain to be intervals of the real numbers. If for the neighbourhood Σ of $f(a)$ we take an interval

$$\{y : f(a) - \varepsilon < y < f(a) + \varepsilon\},$$

then we require $f^{-1}(\Sigma)$ to contain a neighbourhood of a, which can always be taken to be a neighbourhood of the form

$$\{x : a - \delta < x < a + \delta\}.$$

That is, given $\varepsilon > 0$ we can find δ such that

$$|x - a| < \delta \Rightarrow |f(x) - f(a)| < \varepsilon.$$

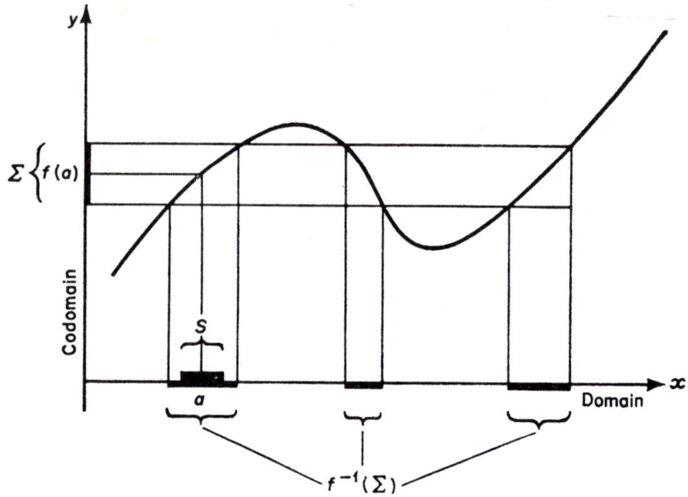

However, the advantage of leaving the definition in the 'neighbourhood' shape is that the same form of words can be used later on for functions of complex variables, functions of two real variables, and so on. The only difference is that the neighbourhoods look rather different in the diagram.

4.17 Mean Value Theorems for Higher Derivatives

The well-known theorem that if $f(t)$ possesses a second derivative, then there is a number ξ between $x-h$ and $x+h$ such that

$$\{f(x+h)-2f(x)+f(x-h)\}/h^2 = f''(\xi),$$

may be proved by considering the auxiliary function

$$\phi(t) = f(t)-At-Bt^2.$$

For A and B may be chosen so that $\phi(x-h) = \phi(x) = \phi(x+h)$, and then, by Rolle's theorem, there exist numbers ξ_1 and ξ_2, $x-h < \xi_1 < x < \xi_2 < x+h$ (supposing h positive), such that

$$\phi'(\xi_1) = \phi'(\xi_2) = 0.$$

Hence, again by Rolle's theorem, there exists a number, ξ, $\xi_1 < \xi < \xi_2$, such that $\phi''(\xi) = 0$. But $\phi''(\xi) = f''(\xi) - 2B$, and from the defining equations for A and B we easily find that

$$2h^2 B = f(x+h) - 2f(x) + f(x-h),$$

which establishes the result.

By considering

$$\phi(t) = f(t) - A_1 t - A_2 t^2 - \ldots - A_n t^n$$

in the same way, the corresponding result for the nth derivative may be obtained. The coefficient A_n is, of course, the only one that need be determined.

4.18 Reciprocals as an Introduction to Iteration

The usual introduction to the important process of iteration is by way of Newton's formula for the solution of $f(x) = 0$; $x_{n+1} = x_n - f(x_n)/g$, where g is the gradient of $f(x)$ in the neighbourhood of the zero. (The formula $x_{n+1} = x_n - f(x_n)/f'(x_n)$ is usually more tedious in computation and may even be less rapidly convergent.) There is, however, something to be said for introducing iteration by means of a process which requires less computational skill, and the formula $x_{n+1} = x_n(2 - ax_n)$ converging to $x = 1/a$ seems admirably suitable. 'Division by multiplication' quickly arouses interest.

The following is a possible approach. At any stage of the process $1 - ax_n$ is to be small; we then want to have $1 - ax_{n+1}$ very much smaller. The suggestion $1 - ax_{n+1} = (1 - ax_n)^2$ seems an obvious choice and leads rapidly to the formula. The convergence rate is thus known; we may expect to double the number of significant figures at each stage. For example, taking $a = 3$, $x_1 = \cdot 3$ we obtain $x_2 = \cdot 33$, $x_3 = \cdot 3333$, and so on, x_n having 2^{n-1} digits of the recurring decimal.

Similar considerations can be used to lead to the iterative process for square root. If we want $x_{n+1} - \sqrt{a}$ to be of the order $(x_n - \sqrt{a})^2$, we can write $x_{n+1} - \sqrt{a} = k(x_n - \sqrt{a})^2 = k(x_n{}^2 + a) - 2kx_n\sqrt{a}$. The formula will not involve the unknown $\sqrt{a}$ only if $2kx_n = 1$, and this leads to $x_{n+1} = \frac{1}{2}(x_n + a/x_n)$.

4.19 Some Irrational Number Approximations

I π (using Simpson's rule)

Inscribe a regular hexagon in a circle of radius 12. Taking ordinates a, b, c, d, e, f, g at intervals of one unit as shown in the figure, Simpson's Rule gives the approximation

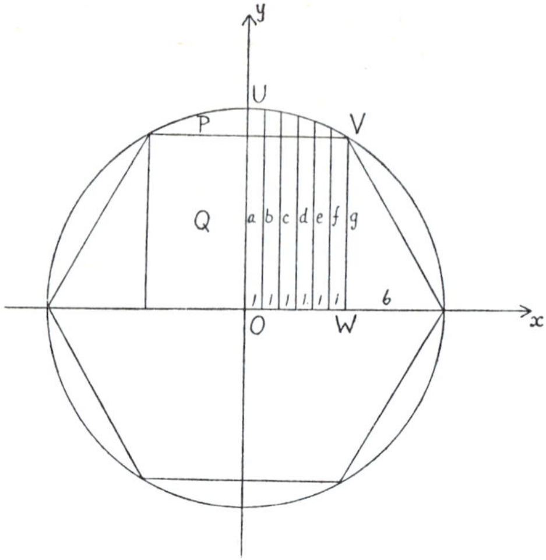

$$\text{area UVWO} = P+Q = \tfrac{1}{3}(a+4b+2c+4d+2e+4f+g). \qquad (1)$$

Now the whole circle equals $12P+$ hexagon, so that

$$144\pi = 12P+6Q = 12(P+Q)-6Q. \qquad (2)$$

Using the fact that $Q = 6g$, equations (1) and (2) give immediately

$$\pi = \tfrac{1}{36}(a+4b+2c+4d+2e+4f-8g),$$

or

$$\pi = \tfrac{1}{36}(12+4\sqrt{143}+2\sqrt{140}+4\sqrt{135}+2\sqrt{128} +4\sqrt{119}-8\sqrt{108}).$$

Using Chamber's *Shorter Six-figure Mathematical Tables* this gives $\pi = 3{\cdot}141\ 583$ compared with the correct figure of $3{\cdot}141\ 593$.

The method extends immediately to circles of greater (or smaller) radius. The value $r = 8$ gives $\pi = \frac{1}{16}(8+4\sqrt{63}+2\sqrt{60}+4\sqrt{55}-5\sqrt{48}) = 3{\cdot}141\ 545$. The value $r = 16$ gives $\pi = 3{\cdot}141\ 590$. For $r = 4$ we get the remarkably simple and accurate result $\pi = 1+\sqrt{15}-\sqrt{3} = 3{\cdot}141$, using 4 figure tables.

II $\sqrt{2}$

Since $0 < \sqrt{2}-1 < 1$, $(\sqrt{2}-1)^n \to 0$ as $n \to \infty$.

If $$(\sqrt{2}-1)^n = (-1)^n(p_n - q_n\sqrt{2}),$$
then $$(\sqrt{2}-1)^{n+1} = (-1)^{n+1}[(1-\sqrt{2})(p_n - q_n\sqrt{2})];$$
so that $$p_{n+1} - q_{n+1}\sqrt{2} = (1-\sqrt{2})(p_n - q_n\sqrt{2}).$$

Hence
$$p_{n+1} = p_n + 2q_n, \quad q_{n+1} = p_n + q_n, \quad \text{and} \quad p_n/q_n \to \sqrt{2}.$$

Starting with $p_1 = q_1 = 1$, we obtain in succession
$$(\sqrt{2}-1)^2 = 3-2\sqrt{2}, \quad (\sqrt{2}-1)^3 = 5\sqrt{2}-7,$$
$$(\sqrt{2}-1)^4 = 17-12\sqrt{2}\ldots$$
which gives the successive approximations for $\sqrt{2}$ as
$$\frac{3}{2}, \frac{7}{5}, \frac{17}{12}, \frac{41}{29}, \ldots$$

The last is 1·414, correct to 4 SF.

III e

If $$I_n = \int_1^e (\log x)^n\,dx = \int_0^1 u^n e^u\,du,$$
then $$0 < I_n < e\int_0^1 u^n\,du = \frac{e}{n+1}, \quad \text{and} \quad I_n = e - nI_{n-1}.$$

Hence $I_n \to 0$ as $n \to \infty$, and we obtain in succession
$$I_0 = e-1, \quad I_1 = 1, \quad I_2 = e-2, \quad I_3 = 6-2e, \quad I_4 = 9e-24, \ldots$$

If $$I_n = (-1)^n(p_n e - n!),$$
then $$p_n = np_{n-1} + (-1)^n,$$
$n!/p_n \to e$, and the successive approximations are
$$2, \frac{6}{2}, \frac{24}{9}, \frac{120}{44}, \frac{720}{265}, \frac{5040}{1854}, \ldots$$

The last is 2·718, correct to 4 SF.

4.20 Miscellaneous Methods of Finding Approximate Values of Logarithms

I Euler in his *Elements of Algebra* gives the following simple method for calculating logarithms. He takes $\log_{10} 2$ as an example.

Let $\log_{10} 2 = x$, so that $10^x = 2$.

$$\begin{aligned}
10^0 &= 1, & 10^1 &= 10; & &\text{hence} & 0 &< x < 1.\\
(10^{1/2})^2 &= 10; & 2^2 &= 4; & &\text{hence} & 0 &< x < \tfrac{1}{2}.\\
(10^{1/3})^3 &= 10, & 2^3 &= 8; & &\text{hence} & 0 &< x < \tfrac{1}{3}.\\
(10^{1/4})^4 &= 10, & 2^4 &= 16; & &\text{hence} & \tfrac{1}{4} &< x < \tfrac{1}{3}.
\end{aligned}$$

Using, though he does not say so, the result that $(a+c)/(b+d)$ lies between a/b and c/d, he proceeds:

$$\begin{aligned}
(10^{2/7})^7 &= 100, & 2^7 &= 128; & &\text{hence} & \tfrac{2}{7} &< x < \tfrac{1}{3}.\\
(10^{3/10})^{10} &= 1000, & 2^{10} &= 1024; & &\text{hence} & \tfrac{3}{10} &< x < \tfrac{1}{3},
\end{aligned}$$

and so on, to any required degree of accuracy.

II It will be found that the formula

$$\log_e t = 6(t-1)/(t+1+4\sqrt{t})$$

reproduces the Napierian logarithm of t with almost five figure accuracy over the extensive range $1 < t < 2$.

The formula is easily established by applying Simpson's rule of approximate integration to the integral $\int_0^1 t^x\,dx$.

The high accuracy is probably due to the fact that the remainder in Simpson's rule, expressed as an expansion in ascending powers of the difference operator over a unit interval, begins with a multiple of $\Delta^4 - \frac{3}{2}\Delta^5$ which vanishes for the function $f(x) = (\frac{5}{3})^x$; noting $1 < \frac{5}{3} < 2$.

III $$\log_e(1+x) = \log_e\left\{\left(1+\frac{x}{2+x}\right)\Big/\left(1-\frac{x}{2+x}\right)\right\}.$$

If we write $x/(2+x) = z$,

$$\log_e \{(1+z)/(1-z)\} = 2\{z+z^3/3+z^5/5+\ldots\},$$

and

$$6z/(3-z^3) = 2\{z+z^3/3+z^5/9+\ldots\},$$

and the difference is $8z^5/45+\ldots$.

Hence $\log_a(1+x) = 6z/\{(3-z^2)\log_e a\}$, with an error of about $z^5/13$, when $a = 10$.

Take, for example, $\log_{10}2$. We have $\log_e(1+\frac{1}{4}) = 27/121$, and hence

$$\log_{10}(1+\tfrac{1}{4}) = \cdot 4342945 \times 27/121$$
$$= \cdot 0969087.$$

Add error, $$= \cdot 0000013.$$

Thus $$\log_{10}(1+\tfrac{1}{4}) = \cdot 0969100,$$

so that $$\log_{10}8 = 1 - \cdot 0969100$$
$$= \cdot 9030900,$$

which gives $$\log_{10}2 = \cdot 3010300.$$

Had we first found $\log 1{\cdot}024$, the answer would have been correct to 12 figures.

4.21 Radius of Curvature in Pedal Coordinates

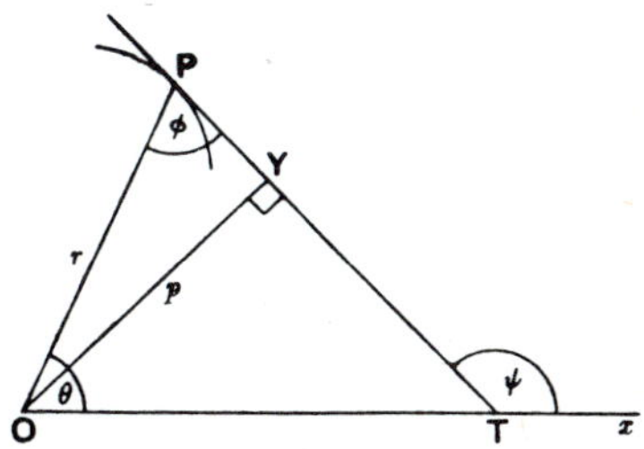

$$\psi = \theta + \phi.$$

Thus $$\kappa = \frac{d\psi}{ds} = \frac{d\theta}{ds} + \frac{d\phi}{ds}. \tag{1}$$

Now

$$\frac{d\theta}{ds} = \frac{1}{r}\left(r\frac{d\theta}{ds}\right) = \frac{\sin\phi}{r}$$

and

$$\frac{d\phi}{ds} = \frac{d\phi}{dr}\cdot\frac{dr}{ds} = \frac{d\phi}{dr}\cdot\cos\phi.$$

Hence from (1)

$$\kappa = \frac{\sin\phi}{r} + \frac{d\phi}{dr}\cos\phi$$

$$= \frac{1}{r}\left(\sin\phi + \frac{d\phi}{dr} \,.\, r\cos\phi\right)$$

$$= \frac{1}{r}\frac{d}{dr}(r\sin\phi) = \frac{1}{r}\frac{dp}{dr},$$

and

$$\rho = r\,dr/dp.$$

4.22 A Source of Error in Solving Differential Equations

This note illustrates an erroneous method for finding a particular integral of a differential equation, and indicates how the mistake may be avoided. For example,

$$(D^2+2D+5)y = 4e^{-x}.$$

Take a particular integral

$$y = pxe^{-x};$$

then

$$Dy = p(-x+1)e^{-x},$$

and

$$D^2y = p(x-2)e^{-x},$$

so

$$4pxe^{-x} \equiv 4e^{-x},$$

$$p = 1/x$$

and

$$y = e^{-x}.$$

This is the correct answer, but the method is faulty, because p has been assumed constant, and then turns out to depend on x. Of course, the wrong form for a particular integral was chosen originally, probably because the complementary function had already been found as $e^{-x}(A\cos 2x + B\sin 2x)$, and the student had the mistaken idea that, since this contains the factor e^{-x}, the particular integral needs an extra factor x.

This mistake is liable to occur whenever we have a differential equation of the form

$$(D^2 - 2\alpha D + \beta)y = ce^{\alpha x},$$

and $\beta > \alpha^2$. It would not occur if one adopted the method of substituting $y = e^{\alpha x}z$, where z is a new variable.

4.23 Connecting the Forms of the Solutions of Differential Equations

The general solution of the equation

$$(D-a)(D-a-h)y = 0$$

is

$$y = Ae^{ax} + Be^{(a+h)x}$$

unless $h = 0$, when it is $y = (A+Bx)e^{ax}$. Students are often puzzled by the fact that substitution of $h = 0$ in the first solution does not produce the second. The following procedure may help to show the connexion.

Consider the solution (when $h \neq 0$) in the form

$$y = C\left[e^{(a+h)z} - e^{ax}\right] + De^{ax},$$

which is clearly general. Then, if C is $O(1/h)$, we may put $C = E/h$, and

$$y = De^{ax} + Ee^{ax}\left[\frac{e^{hx}-1}{h}\right].$$

Since $\lim\limits_{h\to 0}\left[\frac{e^{hx}-1}{h}\right] = x$, this approaches $y = (D+Ex)e^{ax}$ as $h \to 0$.

REFERENCES

4.1. *An Area Property of the Curve $y = x^n$*
P. G. Scott — C 91, Vol. 46, No. 357, p. 236

4.2. *Extending the Trapezial Rule for Area*
D. W. Allen — C176, Vol. 52, No. 382, p. 375

4.3. *Functions, Differences and the Fundamental Theorem of the Calculus*
A. D. Woodall — C217, Vol. 54, No. 389, p. 277

4.4. *How Arbitrary can a Constant of Integration be?*
J. D. Griffiths — C287, Vol. 57, No. 401, p. 200

4.5.	*Integration of Inverse Functions*	
	N. Anderson	C205, Vol. 54, No. 387, p. 52
4.6.	*Some Integration Techniques*	
	B. K. Sen-Gupta	C288, Vol. 57, No. 401, p. 201
	S. R. N. Smith	C219, Vol. 54, No. 389, p. 281
	S. Richardson	1170, Vol. 19, No. 236, p. 369
	R. W. Payne	C 95, Vol. 46, No. 357, p. 240
	A. K. Srinivasan	2043, Vol. 33, No. 303, p. 36
4.7.	*An Old Problem: a Donkey Tethered to the Edge of a Circular Field*	
	J. Pedoe	1477, Vol. 24, No. 261, p. 286
4.8.	*Investigating Stationary Points*	
	E. B. Spratt	C119, Vol. 48, No. 366, p. 426
	'I. N. Cog'	C136, Vol. 50, No. 372, p. 153
4.9.	*The Gradient of the Sine Curve*	
	D. A. Quadling	C104, Vol. 47, No. 360, p. 150
	H. A. Thurston	C112, Vol. 48, No. 363, p. 74
4.10.	*Differentiation of Logarithms*	
	E. J. Hopkins	C106, Vol. 47, No. 360, p. 153
4.11.	*Another Approach to the Logarithmic and Exponential Functions*	
	A. G. Sillitto	2488, Vol. 39, No. 327, p. 54
4.12.	*Continuous Interest and the Exponential Function*	
	W. J. Fearns	2777, Vol. 42, No. 341, p. 225
4.13.	*Limit of* $\sin\theta/\theta$ *as* $\theta \to 0$	
	F. Bowman	2029, Vol. 32, No. 302, p. 293
4.14.	*The Convergence of* x^n	
	R. L. Goodstein	2425, Vol. 38, No. 324, p. 140
4.15.	*The Harmonic Series*	
	A. J. B. Ward	C216, Vol. 54, No. 389, p. 277
	G. W. Brewster	2365, Vol. 37, No. 322, p. 279
4.16.	*A Topological Approach to Continuity*	
	D. A. Quadling	C183, Vol. 53, No. 383, p. 56
4.17.	*Mean Value Theorems for Higher Derivatives*	
	D. A. T. Wallace	2207, Vol. 35, No. 312, p. 117
4.18.	*Reciprocals as an Introduction to Iteration*	
	H. M. Cundy	C 75, Vol. 45, No. 354, p. 333
4.19.	*Some Irrational Number Approximations*	
	F. Garside	C218, Vol. 54, No. 389, p. 280
	R. Haydock	C120, Vol. 48, No. 366, p. 426
4.20.	*Miscellaneous Methods of Finding Approximate Values of Logarithms*	
	H. F. Sandman	1865, Vol. 30, No. 288, p. 34
	S. J. Tupper	2791, Vol. 42, No. 342, p. 286
	R. H. Birch	1802, Vol. 29, No. 283, p. 21
4.21.	*Radius of Curvature in Pedal Coordinates*	
	F. J. Tongue	1233, Vol. 21, No. 243, p. 154
4.22.	*A Source of Error in Solving Differential Equations*	
	E. J. F. Primrose	C239, Vol. 55, No. 391, p. 51
4.23.	*Connecting the Forms of the Solutions of Differential Equations*	
	T. J. Randall	C145, Vol. 50, No. 374, p. 386

5 Geometry

To many people geometry is a formal subject involving the mystic ritual of proof. However, the opening notes of this section illustrate the broad range of the subject from puzzle to traditional theorem.

5.1 Directed Doodles

Chance doodling led to this happy combination of directed numbers and symmetrical patterns, which in practice has created much interest among 12-year olds, and gives plenty of scope for discovery.

A sequence of integers is chosen, with sum zero. This sequence is taken as representing X displacements, and is interleaved with the same sequence for Y displacements. For example, the sequence

$$+2, \quad -1, \quad +4, \quad -2, \quad +1, \quad -4$$

may be interleaved with itself as

X	$+2$	-1	$+4$	-2	$+1$	-4
Y	$+4$	-2	$+1$	-4	$+2$	-1

This is then read as a sequence of displacements, alternately of X and of Y. Thus, starting from an arbitrary point, take $+2$ in X direction, $+4$ in Y direction, and so on. The doodle generated returns to the starting point since the sum of each row is zero. The result is in the figure.

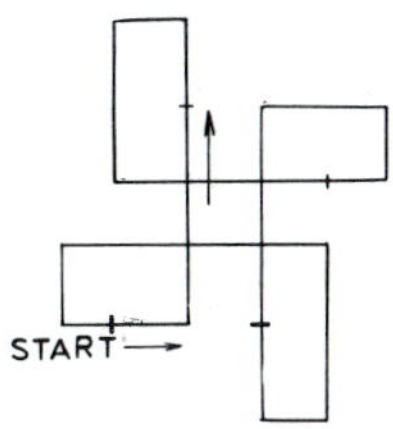

The doodle is then varied by altering the Y-series relative to the X-series, by moving along the relative positions, or by reversing the Y-series. This one basic series gives five rewarding doodles.

Try your hand at the doodle-bug, starting from $+2$, -3, $+4$, -2, $+3$, -4.

If you should think this too simple, try finding the conditions that the resulting doodle should have rotational symmetry, or line symmetry, or both!

5.2 Pythagoras' Theorem†

If ABC has a right angle at C and angles x, y at A, B, and if CDE, placed as shown in Fig. 1, is congruent to BCA, then

$$\angle CKA = 180° - (x+y) = 90°.$$

FIG. 1

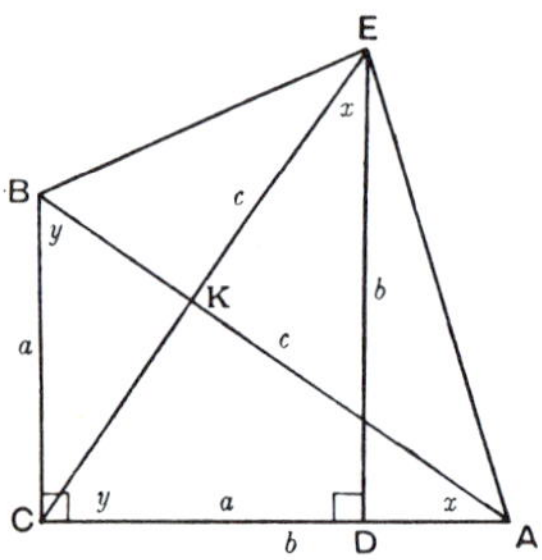

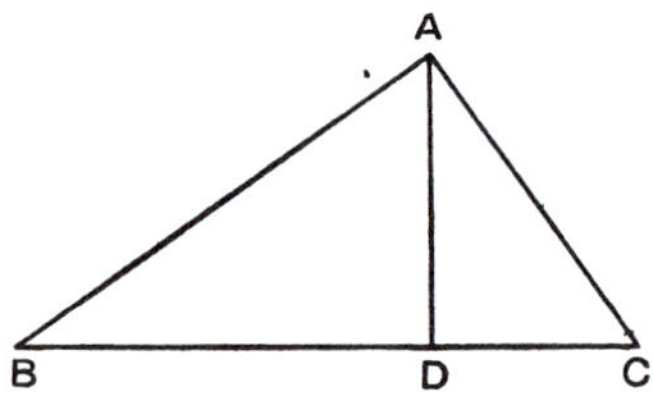

FIG. 2

Then area $ACBE$ = area CBE + area ACE, or

$$\tfrac{1}{2}c^2 = \tfrac{1}{2}a^2 + \tfrac{1}{2}b^2.$$

Alternatively, from the similar triangles ABC, DAC, DBA (Fig. 2)

$$\frac{\triangle ABC}{a^2} = \frac{\triangle DAC}{b^2} = \frac{\triangle DBA}{c^2}$$

$$= \frac{\triangle DAC + \triangle DBA}{b^2 + c^2}$$

$$= \frac{\triangle ABC}{b^2 + c^2}.$$

Hence $$a^2 = b^2 + c^2.$$

† See also Glenn, W. H. and Johnson, D. A., *Number Patterns*, Exploring Mathematics on your own, 3, John Murray, London.

Apollonius' theorem says that if M is the mid-point of BC in a triangle ABC, then $AB^2+AC^2 = 2BM^2+2AM^2$. This theorem leads to the following remarks.

5.3 A Curious Triangle

If a triangle be similar to the triangle formed by its medians, prove that the sum of the squares on two sides of the triangle is equal to twice the square on the third side.

The proof is simple. If $a > b > c$ and the medians AD, BE, CF are of lengths l, m, n, then $n > m > l$, $\frac{l}{c} = \frac{m}{b} = \frac{n}{a}$ and Apollonius' theorem gives

$$4l^2 = 2b^2+2c^2-a^2 = \lambda c^2$$
$$4m^2 = 2c^2+2a^2-b^2 = \lambda b^2$$
$$4n^2 = 2a^2+2b^2-c^2 = \lambda a^2,$$

from which it follows that, by adding, $\lambda = 3$ and $a^2+c^2 = 2b^2$.

5.4 Converses

So many converses of elementary theorems are true that pupils are apt to take a converse for granted (possibly with a twinge of mathematical conscience) and to regard demand for proof as pedantry.

The writer considers that the theorem of Apollonius (quoted above, 5.2) is an excellent elementary example of a theorem whose converse is not true; there are in general two positions of M on BC for which $AB^2+AC^2 = 2BM^2+2AM^2$, one being the mid-point. Otherwise we might have had

Suinollopa's Theorem: Every straight line is bisected by its end points. For, taking P (preferably) not on XY such that $PX = PY$, we have $PX^2+PY^2 = 2XX^2+2PX^2$; and hence X is the mid-point of XY.

But here is a converse which does hold:

The following proof was demonstrated by a Dutch officer to another prisoner of war in a camp in Java. It does not appear to be well known in this country.

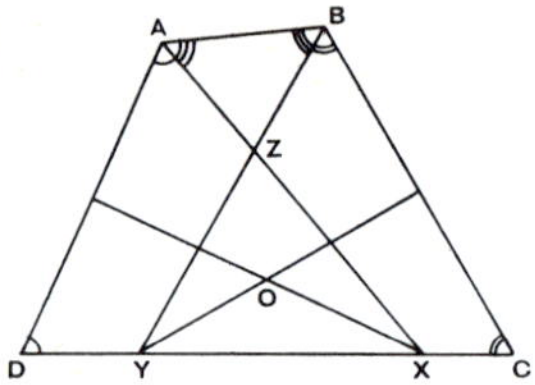

Given: a quadrilateral $ABCD$ with a pair of opposite angles supplementary.

To prove: that $ABCD$ is cyclic.

Construction: Draw AX, BY so that $\angle XAD = \angle ADX$, $\angle YBC = \angle BCY$, to meet DC in X, Y respectively. Let AX, BY meet in Z.

Proof: Since $\angle DAB + \angle BCY = \angle ABC + \angle ADX$

$$\angle DAB - \angle ADX = \angle ABC - \angle BCY$$

Thus by construction $\angle ZAB = \angle ZBA$.

Hence the three triangles XAD, YBC, ZAB are all isosceles. It follows that the perpendicular bisectors of AD, BC pass through X and Y respectively. Let them meet in O. Since the triangles XAD, YBC are isosceles, XO and YO must be the bisectors of the angles X and Y, in the triangle XYZ.

Hence OZ must be the bisector of the angle Z, and therefore OZ produced is the perpendicular bisector of AB, since the triangle ZAB is isosceles.

Hence a circle, centre O, can be drawn to pass through A, B, C, D.

The following give a stock of neat integer-sided figures illustrating some of the classical theorems.

5.5 Integral-sided Geometric Figures

I Just as some people love long computations, so others are fascinated by integral-sided geometric figures. Many may have noticed that

$$25^2 = 625 = 400 + 225 = 20^2 + 15^2$$
$$= 576 + 49 = 24^2 + 7^2.$$

This, of course, means that in a semi-circle having diameter 25 one can fit two right-angled triangles with integral sides.

Now it is also true that

$$(20 \times 7) + (15 \times 24) = 25 \times 20.$$

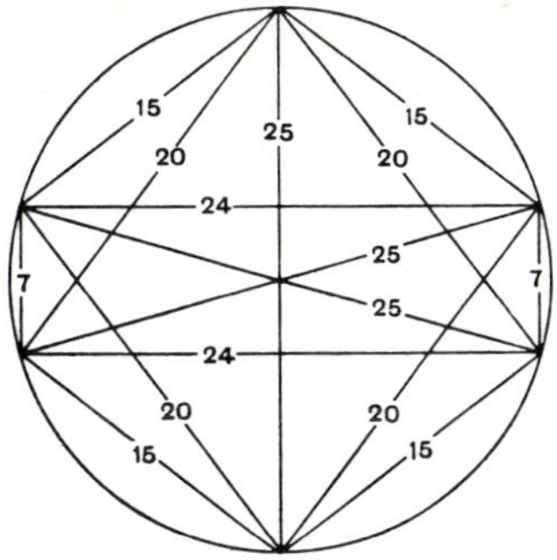

Ptolemy's theorem is suggested, giving a cyclic quadrilateral with one of its diagonals and also one of its sides of 20 units. This leads to an isosceles triangle . . . and by the time one has finished with the figure it is a cyclic hexagon with all sides and diagonals of integral lengths, and quite small integers, too. Meccano strips can be used with effect.

II *Pythagorean triangles.* The general formulae are very well known, but are mentioned for the sake of completeness: $a^2+b^2 = c^2$ if $a = l^2-m^2$, $b = 2lm$, $c = l^2+m^2$ (where to avoid solutions reducible to simpler ones l, m are co-prime and not both odd).

III *Heronian triangles.* These have integral area, and hence rational altitudes and radii of in-, ex- and circumcircles. Numerous applications, e.g. to the s-formula for the area and to easily-checked elementary constructions, will suggest themselves. The expressions $l+m$, $m+n$, $n+l$, where $l+m+n = lmn$ give rational values for the sides, but it is much easier to make Heronian triangles by fitting together Pythagorean triangles with equal sides (adjacent to the right angles). Such triangles, reduced if necessary, will always have one integral altitude.

IV *60° and 120° triangles.* Integral-sided examples of these are useful for making up exercises on the cosine rule. They can be found from the formulae:

If

$$\begin{aligned} a &= 3l^2-2lm-m^2 \\ a' &= 4lm \\ b &= 3l^2+2lm-m^2 (= a+a') \\ c &= 3l^2+m^2, \end{aligned}$$

then:
$$\begin{aligned} c^2 &= a^2+b^2-ab \\ &= a'^2+b^2-a'b \\ &= a^2+a'^2+aa', \end{aligned}$$

so that in $\triangle$'s a, b, c; a', b, c; $\angle C = 60°$, and in $\triangle a, a', c$; $\angle C = 120°$.

V For the theorem of Apollonius, one can take any right-angled triangle obtained by the usual formula for Pythagorean numbers and draw the median from the right angle to the hypotenuse, but this may lead the pupils to think that the theorem only applies to right-angled triangles. What one wants is a formula of the type

$$s^2+t^2 = 2l^2+2m^2,$$

and each solution will give two triangles, since either l or m may be the median or half-base.

Incidentally one must test to see that the figure is possible.

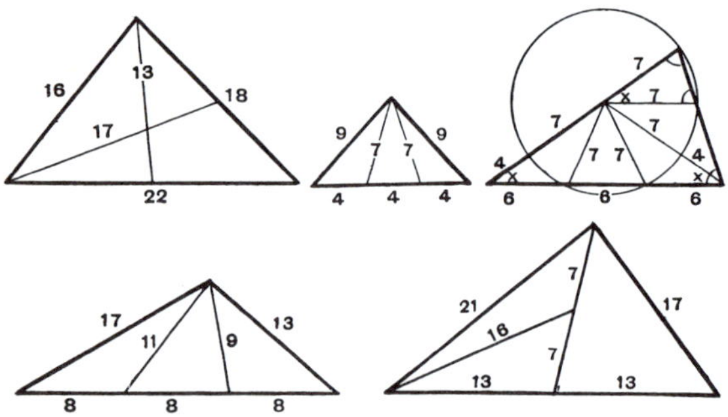

Several formulae for Apollonian numbers can be devised:

$$\begin{aligned} (4a+b)^2+(2a+3b)^2 &= 2(3a+b)^2+2(a+2b)^2; \\ (a^2-b^2)^2+(a^2+7b^2)^2 &= 2(a^2-5b^2)^2+2(4ab)^2; \\ (10a^2-10b^2)^2+(12ab)^2 &= 2(7a^2-b^2)^2+2(7b^2-a^2)^2. \end{aligned}$$

Only the first proves to be of real use, though the other two are of interest (most formulae give flat lines instead of triangles). By noticing if one triangle turns up twice, or if one half-triangle is isosceles or occurs in different cases, one is able to construct compounded figures.

Here is an examination question showing how use may be made of these results:

The longest side of a triangle of sides 13, 17 and 24 cm is trisected; and lines drawn from the points of trisection to the opposite vertex. Find their lengths. (Hint: use the theorem of Apollonius.)

Here are a couple of unusual aspects of circles and rectangles.

5.6 Rectangling the Circles

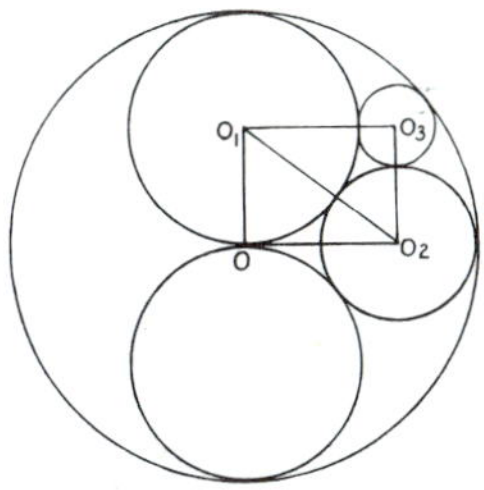

Some readers may be surprised by the following discovery with regard to the above system of circles O, O_1, O_2, O_3 touching one another — that the radii of the circles centres O_3, O_2, O_1 are in the ratio 1 : 2 : 3 and that the triangles OO_1O_2 and $O_1O_2O_3$ are of type '3 : 4 : 5', although this might be well known. The proof is well within the range of an O-level pupil.

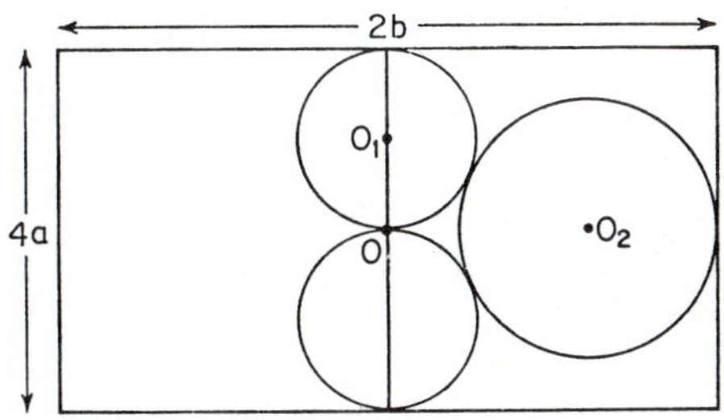

If O is the centre of rectangles in which one pair of opposite sides of length $2b$ touch the given two equal circles of radius a at the extremities of the diameters through O, and O_2 is the centre of a circle touching the equal circles and the side $4a$, and lying within the rectangle, some further simple relationships emerge.

Let t be the ratio $b : a$. Provided t does not exceed $2(1+\sqrt{2})$ — corresponding to the extreme case when circle centre O_2 touches side

$2b$ also — there are two examples of a particular type of right-angled triangle OO_1O_2, which arises when the product of the t values is 2. For example, type $8:15:17$ for $t = 3$ and $\frac{2}{3}$, and type $5:12:13$ for $t = 4$ and $\frac{1}{2}$.

Triangle OO_1O_2 is of type $3:4:5$ when $t = 1$ and 2, that is when the rectangle circumscribes the equal circles and becomes a square respectively.

Also for $t = 1$, the radii of the circles centres O_3, O_2, O_1 are in the ratio $4:9:36$.

5.7 The Maltitudes of a Cyclic Quadrilateral

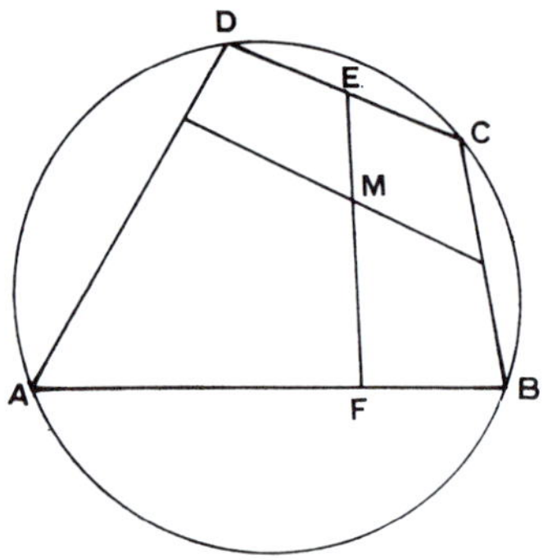

ABCD is a cyclic quadrilateral with centre *O*.

A line through *E* (mid-point *DC*), perpendicular to the opposite side *AB* we call a 'maltitude' (median at one end, altitude at the other). There are six maltitudes and they are concurrent (at *M* say). The centre of gravity of equal weights at *ABCD* lies on *OM* and bisects it, making *OM* rather like an Euler line. The four Simson lines of each vertex with respect to the triangle formed by the other three all pass through *M*.

The four nine-point circles of the triangles formed by any three vertices all pass through *M*. The line joining *A* to the orthocentre of *BCD* goes through *M* and is bisected there, etc.

All these properties are easily proved by elementary geometry.

It would seem that *M* is a rather special point for the quadrilateral.

Returning to a traditional topic, here are some unusual proofs of a famous result.

5.8 The 'Alternate Segment' Theorem

(i) A short while ago I had occasion to bend some pieces of wood to make an edging for a circular flower bed. While doing so the thought occurred to me that the more flexible the wood the more I could bend it and the less number of pieces I would require to make a circle. Automatically following this line of thought, it next occurred to me that the amount of bending would be measured by the angle between the tangents at the ends, and the fraction of the circumference covered by the wood would be measured by the angle subtended by the wood at the centre. I then realised that I had an unusual proof of the 'alternate segment' theorem. Here it is:

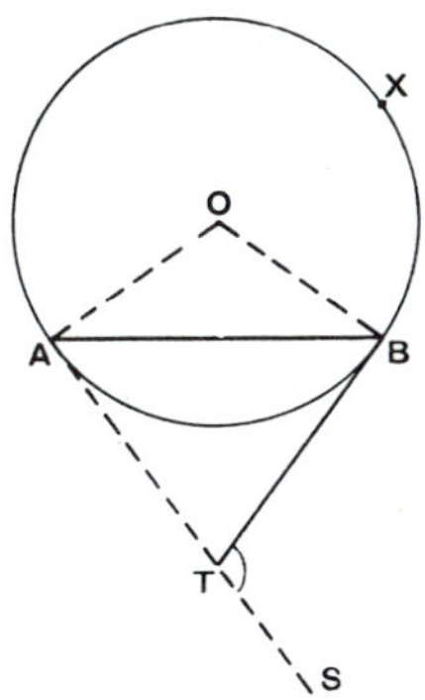

Given a chord AB and a tangent BT.
Join AO, BO, and draw the tangent ATS at A.
Since the angles OAT, OBT are right angles, $AOBT$ is cyclic.
Hence $\angle BTS = \angle AOB$ at centre.
Since $TA = TB$, $\angle ABT = \frac{1}{2} \angle BTS = \frac{1}{2} \angle AOB$
$= \text{any angle in segment } AXB.$

The two theorems involved — properties of cyclic quadrilaterals and equal tangents — are both usually proved before the alternate segment theorem, so there is no logical objection to the proof, but perhaps the interest lies less in the proof itself than in the manner of its discovery.

(ii) The following proof of the alternate segment theorem was produced by a second-form pupil. The class had done little formal geometry but liked 'proving' things. They had spent a short time on symmetry.

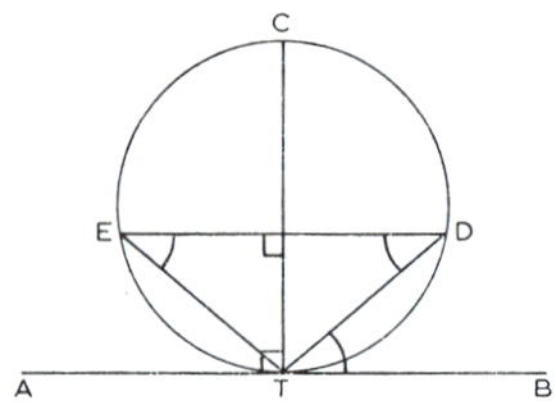

The pupil first drew the circle, the tangent *ATB*, and the diameter *CT*. The figure was then completed as shown, it being symmetrical about *CT*. As a consequence of the symmetry the marked angles are equal, and since $\angle DTB = \angle TED$ it will equal any angle subtended by *DT* in the same segment.

The proof was developed in class and I have no reason to doubt that it is original. I now teach the 'Saltmarsh proof' (after its originator) in preference to the more usual proof.

(iii)

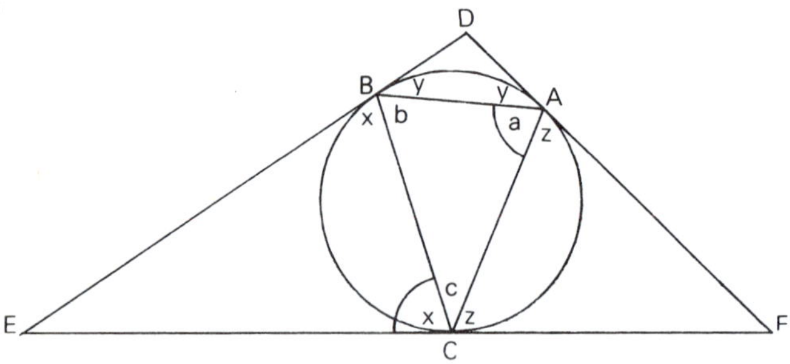

Given: A triangle inscribed in a circle, as in the diagram. (The modification for an obtuse angle is immediate.)
To prove: $\angle x = \angle a$.
Construction: Draw tangents at *A*, *B*, *C*.
Proof: Since tangents from a point to a circle are equal

$$\angle EBC = \angle ECB = \angle x$$
$$\angle DAB = \angle DBA = \angle y$$
$$\angle FCA = \angle FAC = \angle z.$$

Since EF, FD, ED are straight lines,

$$x+y+b = \pi$$
$$x+z+c = \pi$$
$$z+y+a = \pi.$$

Hence $$2x+2y+2z+a+b+c = 3\pi,$$

so that $$x+y+z = \pi.$$

Comparing this with $z+y+a = \pi$, the result follows.

Complex numbers can also be used with benefit for the angle at the centre†:

Proposing rather rashly to a class that complex numbers could be used to provide simple proofs of a number of geometrical theorems, I suggested the well-known angle theorem for a circle as one we might try. Our first attempt proved unexpectedly involved, but subsequent thought led me to this, which is as short as could well be desired. I was sure that it must have appeared somewhere, but could not trace it; so it may be new to others too.

Let t_1, t_2, t_3 be the affixes of three points on the unit circle $|z| = 1$. We wish to prove that the argument of t_2/t_3 is double the argument of $(t_2-t_1)/(t_3-t_1)$; that is, that

$$\begin{aligned}\arg(t_2/t_3) &= 2\arg\{(t_2-t_1)/(t_3-t_1)\}\\ &= \arg\{(t_2-t_1)^2/(t_3-t_1)^2\}.\end{aligned}$$

Now numbers with unit modulus have conjugates equal to their reciprocals. It follows that

$$\begin{aligned}|(t_2-t_1)^2| &= (t_2-t_1)(\overset{*}{t_2}-\overset{*}{t_1})\\ &= (t_2-t_1)\left(\frac{1}{t_2}-\frac{1}{t_1}\right)\\ &= -(t_2-t_1)^2/t_2t_1,\end{aligned}$$

and this is a real number; so similarly is $-(t_3-t_1)^2/t_3t_1$. Dividing, we have that

$$\frac{(t_2-t_1)^2}{(t_3-t_1)^2}\cdot\frac{t_3}{t_2}$$

is real, from which the result is immediate.

† Zwikker, C., *Advanced Plane Geometry*, North Holland, 1950.

Returning to less formal topics, the following are some of the tessellations which have appeared over the years, together with the grids which make their construction easier.

5.9 Mathematical Tiles†

I

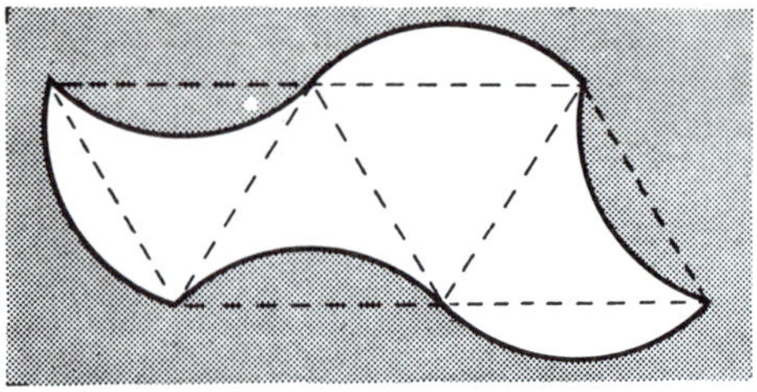

II

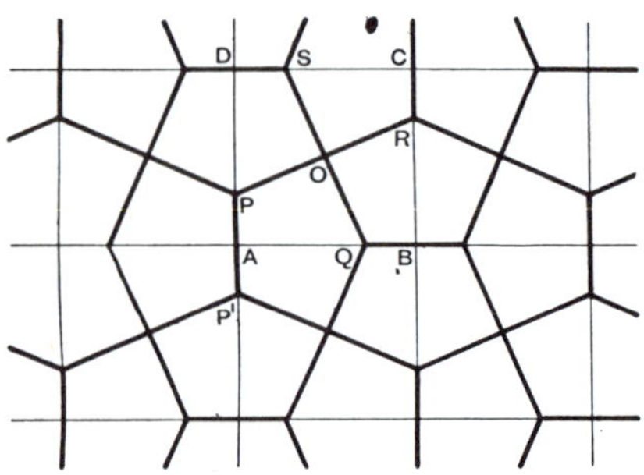

† Tessellations and geometric transformations are used with startling imagination in: *The graphic work of M. C. Escher*, Pan/Ballantine, 1972.

This pentagonal tessellation can be constructed easily on squared paper. If *ABCD* is a typical square, points *P*, *Q*, *R*, *S* on *DA*, *AB*, *BC*, *CD* such that $PA = QB = RC = SD$ are marked; *PR*, *QS* of course are at right angles and pass through the centre *O* of the square.

A simple problem is to find the position of *P* which makes the pentagon equilateral.

III Are there any others besides myself who had not realised that the surprising 'tessellation' property of irregular quadrilaterals becomes obvious when one draws the parallelograms formed by the mid-points of the sides?†

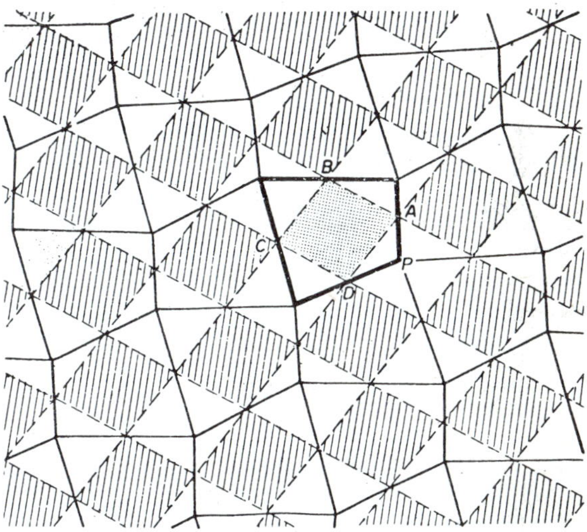

Other geometric puzzles and constructions provide a variety of 'things to do' in the class-room:

5.10 Constructing an Angle of 30°

In setting out to construct an angle of 30°, the following attempt was made by a third former. With centre *O* an arc was drawn. Then

† The relation of this configuration to the symmetry group is given in: Coxeter, H. S. M., *Introduction to Geometry*, John Wiley, 1969.

memory failed, and the second arc with the same radius was drawn with its centre at an arbitrary point P on the line OA to cut the first arc at Q. Q was then used as centre to complete the construction. To everyone's surprise, $\angle ROA$ turned out to be 30° as required. (See Fig. 1).

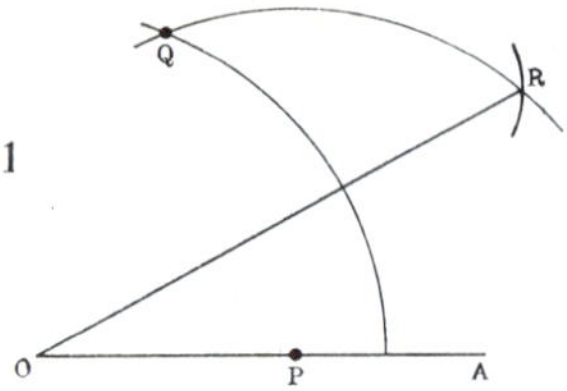

FIG. 1

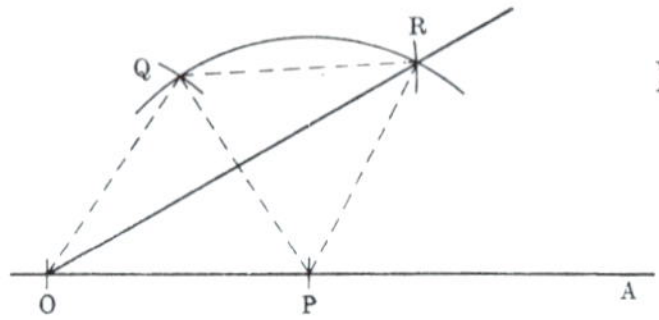

FIG. 2

The proof is simple: a circle, centre Q and radius QO, contains P and R. Hence $\angle ROP = \frac{1}{2} \angle RQP = 30°$.

We may indeed replace the equilateral triangle PQR by any isosceles triangle with $PQ = QR$ and $\angle PQR = 2\alpha$; in this case $\angle ROA = \alpha$.

The four rods OQ, QP, PR, QR form a linkage; if O is fixed and P is constrained to move along the line OA, then R moves along a line making an angle of α with OA. The various positions taken by the triangle PQR give rise to a number of interesting situations, both geometrical and mechanical. For example, if OQ rotates with uniform angular velocity about O, both P and R describe simple harmonic motions with O as centre. (See Fig. 2).

5.11 Uses of a Geometric Puzzle

The form of puzzle of which examples are given here can be used to show the relation between linear dimensions and areas of similar figures.

It is well known that four squares, triangles or parallelograms may be fitted together to make one figure similar to but larger than the original. Puzzle makers and others probably also know of the Figs. 1 and 2.

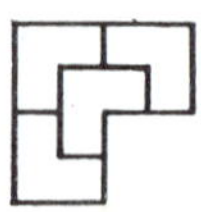

FIG. 1

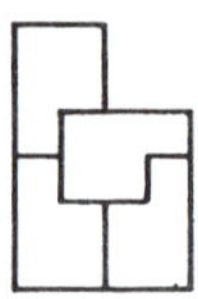

FIG. 2

Figs. 3–7 give some others less well known; that in Fig. 7 always seems to me to be mean. Figs. 3 and 6 are Figs. 1 and 5 bisected.

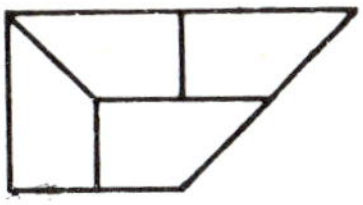

FIG. 3

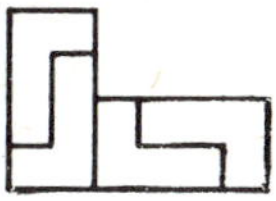

FIG. 4

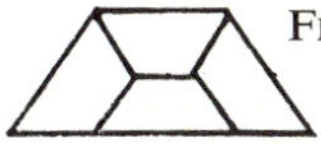

FIG. 5

A class can be shown a few of these and then asked to find others. The problem may, if desired, be extended to solid figures. The difficulty is to find a really pleasing solid corresponding to either Fig. 2 or Fig. 4: a prism with either of these as base will, of course, satisfy the condition required, but this seems more like using thick cardboard in place of paper than finding really new solutions.

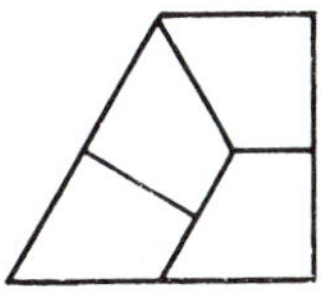

FIG. 6

FIG. 7

There is also the problem of fitting together nine plane figures; thus, for example, Figs. 1 and 5 give the new Figs. 8 and 9, but Figs. 2, 4 and 7 do not seem to extend thus.

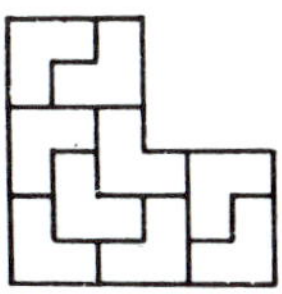

FIG. 8

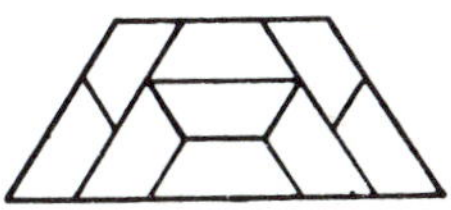

FIG. 9

The figures shown below are curious as 144 of them can be fitted into a similar figure but not any smaller number apparently.

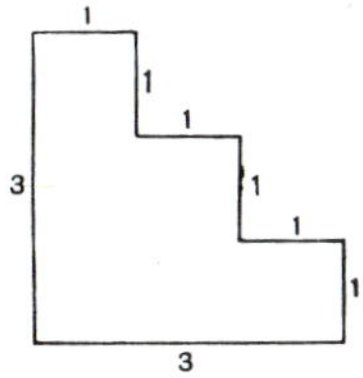

FIG. 10

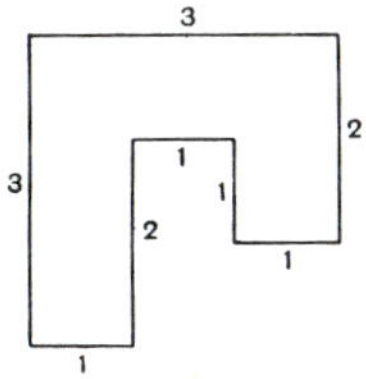

FIG. 11

5.12 The Perpendicular from a Point to a Line

Over the years a very large number of G.C.E. scripts from various examining boards have been marked, but the following construction has been seen on only a very few occasions. The construction seems so simple and elegant that it deserves to be more widely known. Certainly, whenever it has been shown in school there has always been a little chuckle of delight from the class—a sure sign that the elegant simplicity of it has been perceived. A further sign is that once this method has been seen the pupils have steadfastly refused to use any other.

AB is a given straight line and *D* a point on it.

It is required to construct the perpendicular to *AB* at *D*.

Let *P* be any point not on *AB*.

With centre *P* and radius *PD* describe a circle to cut *AB* at *D* and, say, *X*.

Draw *XP* and produce to cut the circle again at *Y*. Draw *DY*.

Then *DY* is the required perpendicular.

This construction is especially useful if the perpendicular at *B* is required.

As one usually deals with the angle properties of a circle before discussing tangent properties, a 'wrong' choice of position for *P*, i.e. one such that *P* would lie on *DY*, has been welcomed—at times even encouraged! Usually, though, it is found that one pupil in each class has asked what would happen if *P* were so chosen. 'Try it and see for yourself' is the inevitable answer. From there on we are into a discussion of tangency.

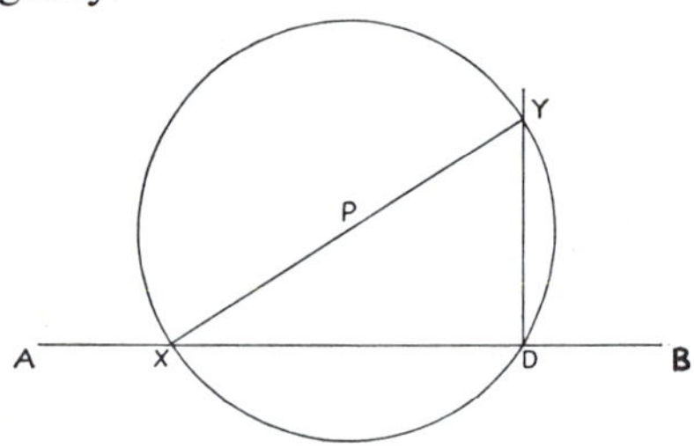

The following gives a neat way of calculating the length of the perpendicular from *P*:

If $P(x_1, y_1)$ is a point not on the line *L*, $ax+by+c = 0$, and $Q(x_2, y_2)$ is the foot of the perpendicular from *P* to *L*, then since *PQ* is perpendicular to $ax+by+c = 0$

$$\frac{y_2-y_1}{x_2-x_1}=\frac{b}{a}$$

or, $$b(x_2-x_1)-a(y_2-y_1)=0. \tag{1}$$

Also since (x_2, y_2) is on the line,

$$\therefore\ ax_2+by_2 = -c$$

or, $$a(x_2-x_1)+b(y_2-y_1) = -(c+ax_1+by_1). \tag{2}$$

Squaring and adding (1) and (2)

$$(a^2+b^2)d^2 = (ax_1+by_1+c)^2.$$

5.13 To Construct a Triangle given the Angle A, and the Sums of Sides c + a, b + a

Construct the triangle APQ, with angle $PAQ = A$, $AP = c+a$, $AQ = b+a$. Cut off equal lengths $PU = QV = k$ from PA, QA respectively. Through U draw a parallel to PQ to meet the circle, centre V and radius k, at X. Then QX produced will meet AP at the required vertex B, and BC is drawn parallel to XV.

5.14 Tangrams and Incommensurables†

(i) Find shapes that can be made from one set (the standard set is shown in Fig. 1), and also from two sets of half the size, i.e. of the same total area.

FIG. 1

FIG. 2

FIG. 3

(ii) Divide a square in some other fashion, and find shapes that can be made from two or from three sets. Two are given in Figs. 2, 3.

(iii) Try to make hexagonal puzzles; one might get something interesting with three angles (30, 60, 90) instead of just the two of original (45 and 90). Figs. 4, 5, 6 show three.

† Tangrams are described further in: Dudeney, H. E., *Five hundred and thirty six puzzles and curious problems*, Scribner, Emblem edition, 1967.

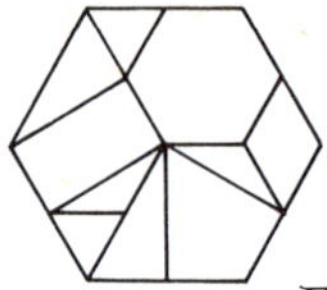

FIG. 4

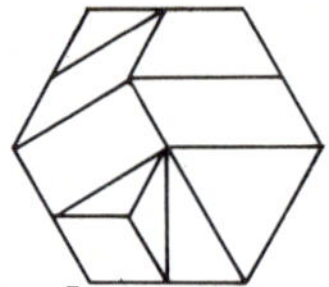

FIG. 5

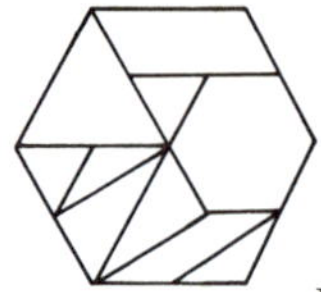

FIG. 6

(iv) Find shapes which can be made from one set in different ways; for example, Fig. 7 can be made from the pieces of Fig. 5 in 123 different ways.

FIG. 7

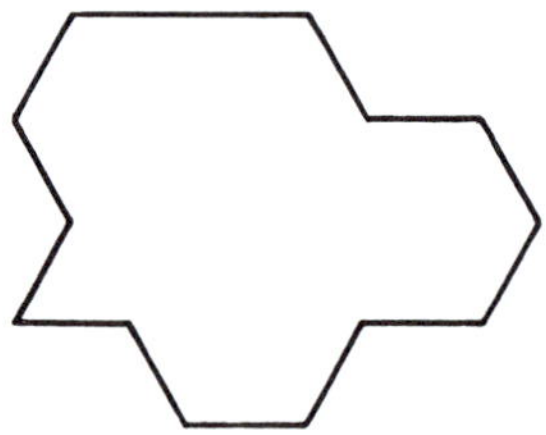

5.15 From the Pentagon to the Dodecahedron

Draw a right angle *AOD*. From *OD* cut off *OB* equal to 2*OA*, and from *BD* cut off *BC*, *CD* each equal to *BA*. Describe a circle on *OD*

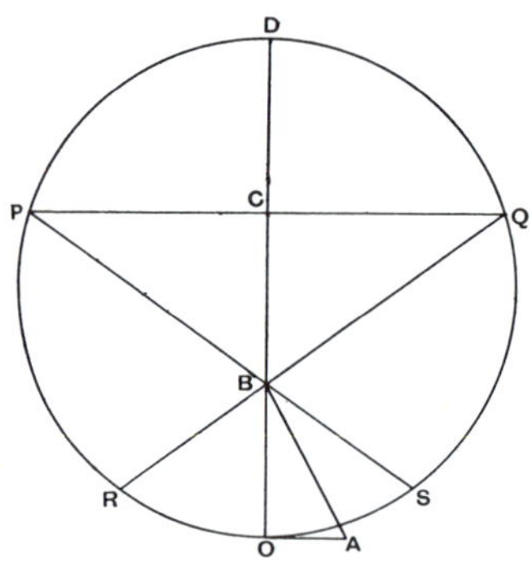

as diameter and draw *PCQ* a chord perpendicular to *OD*. Join *QB* and *PB* and produce them to meet the circumference in *S* and *R*.

Then *DPRSQ* is a regular pentagon.

And having constructed the pentagon, why not use it to make an elegant net for a taste of solid geometry?

The diagonals of a regular pentagon form a pentagram as in Fig. 1 thus forming a pentagon inside the original one.

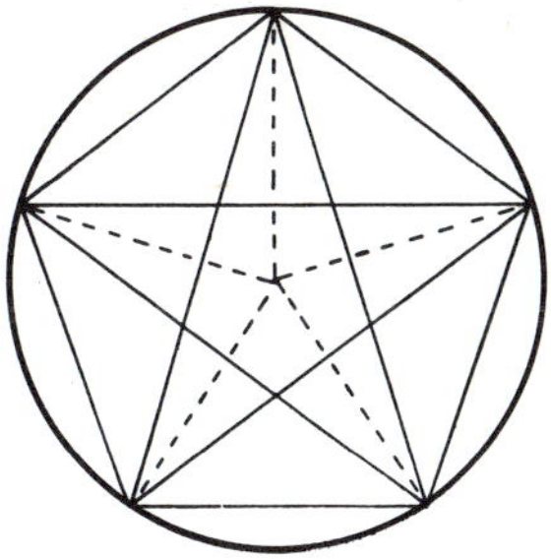

FIG. 1

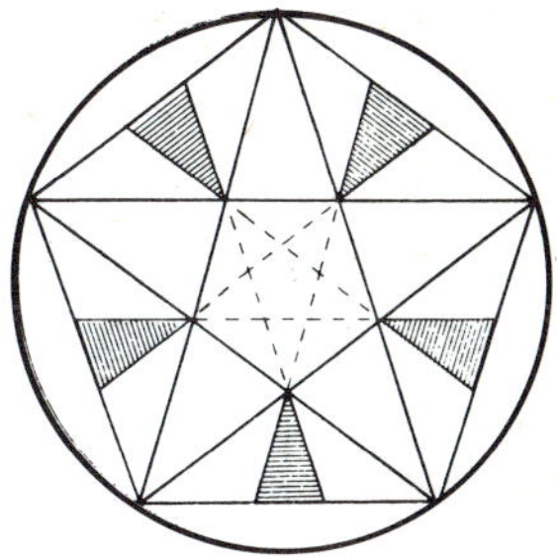

FIG. 2

The diagonals of the inner pentagon are drawn and produced to meet the sides of the outer one as in Fig. 2. If the shaded triangles are then cut away, the resulting figure, consisting of 6 equal pentagons, forms half the net required for the dodecahedron.

The complete net is thus obtained from a small number of initial measurements. If two equal circles are drawn and the vertices of the outer pentagons found by making 5 equal angles at the centre of each circle, the sides can be immediately checked for accuracy with dividers, and the diagrams are then completed by joining points.

This method, in addition to being an easy one to carry out accurately, produces a diagram which is rich in elementary geometry and which can also be used to introduce the idea of the stellation of the solid and which through similar triangles leads to the discovery of a line divided in Golden Section.†

The faithful triangle never fails to produce something fresh:

5.16 A Well-known Geometrical Problem‡

(i) Given that $BD = CE$, prove that $AB = AC$.

Suppose that $\angle ABC \neq \angle ACB$. Without loss of generality, we may take $\angle ABC < \angle ACB$. Draw DF, EG parallel to BC. Then FD falls

† See also Cundy, H. M. and Rollett, A. P., *Mathematical Models*, Oxford University Press, 1961.

‡ Also Coxeter, H. S. M. and Greiter, S. L., *Geometry revisited*, Mathematical Association of America, New Mathematical Library, No. 19.

between EG and BC, for

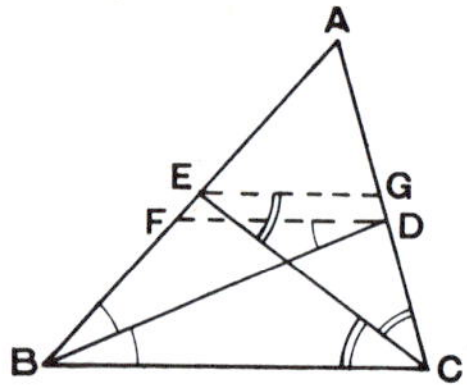

$$CD/DA = BC/BA < BC/CA = BE/EA = CG/GA,$$

whence adding unity to both fractions, $AD > AG$. From the similar triangles AEG, AFD we have

$$EG < FD. \tag{1}$$

Since $\angle GEC > \angle FBD$, we have, by comparing the isosceles triangles GEC, FBD, that

$$EG = GC > FD. \tag{2}$$

This contradicts (1). It follows that $\angle ABC = \angle ACB$; whence $AB = AC$.

(ii)

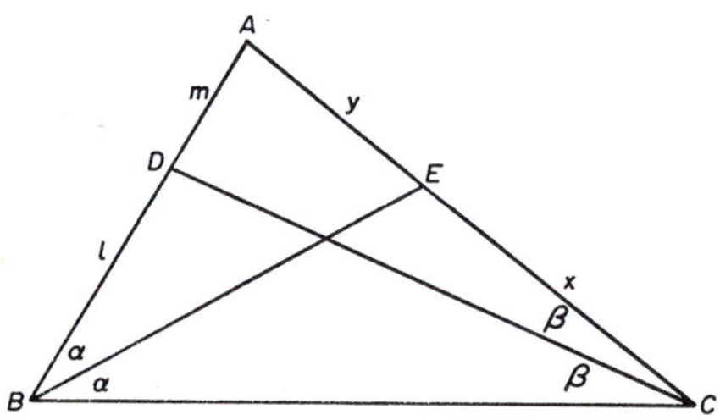

Given that the bisectors BE, CD are equal, to prove that the triangle is isosceles,

$$\triangle ABC \quad \text{gives} \quad \frac{l+m}{\sin 2\beta} = \frac{x+y}{\sin 2\alpha}, \tag{1}$$

$$\triangle\text{s } DCB, BEC \quad \text{give} \quad \frac{l \sin 2\alpha}{\sin \beta} = \frac{x \sin 2\beta}{\sin \alpha}, \tag{2}$$

$$\triangle\text{s } ADC, AEB \quad \text{give} \quad \frac{m}{\sin \beta} = \frac{y}{\sin \alpha}. \tag{3}$$

Substituting for l and m from (2) and (3) in (1) and simplifying:

$$x\frac{\sin\beta}{\sin\alpha}+y\frac{\cos\alpha}{\cos\beta}=x+y.$$

If $\alpha=\beta$, L.H.S. = R.H.S.
If $\alpha<\beta$, L.H.S. > R.H.S. $\qquad \left(\alpha<\frac{\pi}{2},\quad \beta<\frac{\pi}{2}\right)$
If $\alpha>\beta$, L.H.S. < R.H.S.

$\therefore\ \alpha=\beta$ is the only solution.

(iii) The following starts with a construction which depends directly on the equality $BE=CF$, the entire figure being picked up and laid down again with BE coincident with the original position of CF. The assumption that the base angles are unequal then leads to a reductio ad absurdum.

FIG. 1

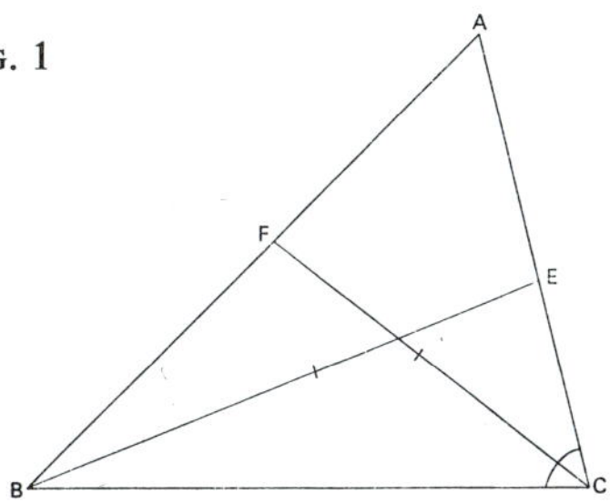

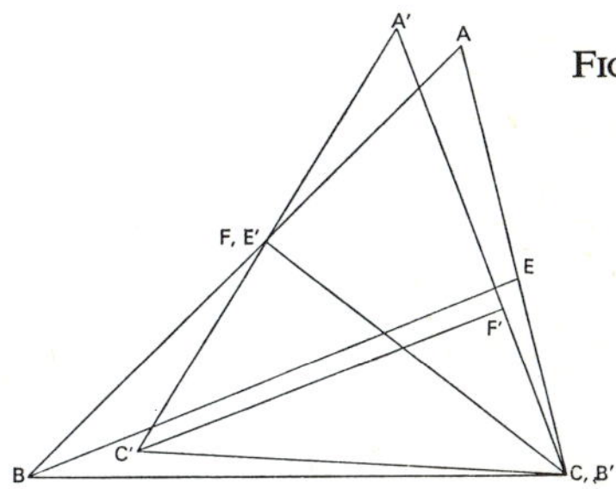

FIG. 2

Suppose that the triangle is not isosceles; then $\angle ABC$ and $\angle ACB$ are not equal, say $\angle ABC<\angle ACB$. Draw the figure $A'B'C'E'F'$ congruent to $ABCEF$, with $B'E'$ coincident with CF and A' near A as shown (Fig. 2). Then $A'C$ lies between AC and FC, and since $\angle FAC=\angle E'A'B'=\angle FA'C$, A' lies outside the triangle FAC. Also, CC' lies between CB and CF, and since $A'FC'$ is a straight line, C' must lie inside the triangle FBC.

From the construction, BE and $C'F'$ make the same angle with CF and are therefore parallel. Hence $C'F'$ is equal and parallel to the base BE of the triangles ABE and CBE and lies entirely within one of them—which is absurd.

But, of course, you all know exactly when two triangles are congruent, don't you?

5.17 When are Triangles Congruent?

Traditionally the answer refers to four cases, namely:

3 sides
2 sides and included angle
1 side and 2 angles
right angle, hypotenuse and 1 side.

'2 sides and the non-included angle' is shown to lead to the 'ambiguous case'. But suppose we are given triangles ABC, PQR such that

$$AB = PQ,$$
$$AC = PR,$$
$$\angle ABC = \angle PQR,$$
$$AC > AB.$$

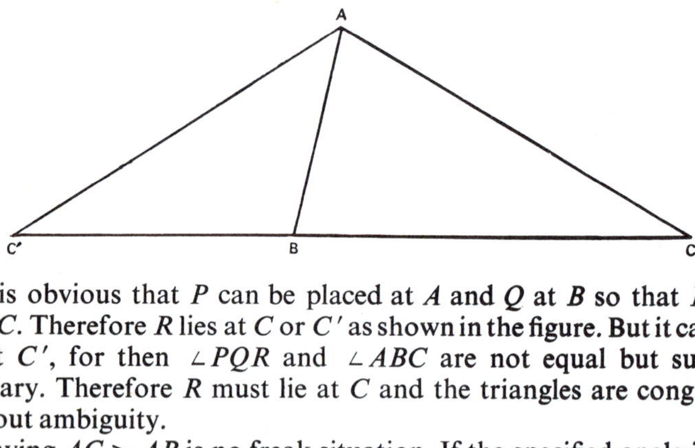

It is obvious that P can be placed at A and Q at B so that R lies on BC. Therefore R lies at C or C' as shown in the figure. But it cannot lie at C', for then $\angle PQR$ and $\angle ABC$ are not equal but supplementary. Therefore R must lie at C and the triangles are congruent without ambiguity.

Having $AC > AB$ is no freak situation. If the specified angle is not included between the specified lines, it is opposite the longer or the shorter one; it will therefore in a sense be opposite the longer side 50% of the time. To reject all non-included cases indiscriminately as ambiguous is therefore not only inexact; it may also be wasteful.

If we accept two triangles as being congruent when the specified angle is opposite the longer side, it is unnecessary to mention the right angle, hypotenuse and one side as a separate case; it is automatically included in the statement above as the right angle is necessarily opposite the longer side. We can therefore reduce congruency to three cases only, namely:

3 sides

1 side and 2 angles

2 sides and 1 angle, provided the angle is not opposite the shorter of the two given sides.

REFERENCES

5.1. *Directed Doodles*
B. D. Price — C145, Vol. 51, No. 375, p. 40

5.2. *Pythagoras' Theorem*
R. Hamilton Dick — 1813, Vol. 29, No. 284, p. 70; 1722, Vol. 28, No. 279, p. 74

5.3. *A Curious Triangle*
A. K. Rajagopol — 2744, Vol. 42, No. 339, p. 40

5.4. *Converses*
R. E. Green — C144, Vol. 51, No. 375, p. 40
N. J. F. Craig — 2111, Vol. 34, No. 307, p. 44

5.5. *Integral-sided Figures*
C. Dudley Langford — 1514, Vol. 25, No. 264, p. 113
A. R. Pargeter — 2895, Vol. 44, No. 348, p. 130
C. Dudley Langford — 1289, Vol. 22, No. 249, p. 187

5.6.	*Rectangling the Circles*	
	D. M. G. Rees	C151, Vol. 51, No. 376, p. 143
	D. W. Allen	C181, Vol. 53, No. 383, p. 54
5.7.	*The Maltitudes of a Cyclic Quadrilateral*	
	R. E. Lester	3020, Vol. 46, No. 356, p. 147
5.8.	*The 'Alternate Segment' Theorem*	
	A. J. Mornard	2032, Vol. 32, No. 302, p. 297
	D. C. Armstrong	C125, Vol. 49, No. 367, p. 73
	Stanley Kravitz	C139, Vol. 50, No. 373, p. 295
	H. Martyn Cundy	C294, Vol. 57, No. 402, p. 333
5.9.	*Mathematical Tiles*	
	H. Martyn Cundy	2647, Vol. 40, No. 334, p. 285
	A. P. Rollett	2530, Vol. 39, No. 329, p. 209
	D. A. Quadling	C182, Vol. 53, No. 383, p. 54
5.10.	*Constructing an Angle of* 30°	
	G. T. Wain	C204, Vol. 54, No. 387, p. 50
5.11.	*Uses of a Geometric Puzzle*	
	C. Dudley Langford	1464, Vol. 24, No. 260, p. 209
	R. Sibson	1465, Vol. 24, No. 260, p. 211
5.12.	*The Perpendicular from a Point to a Line*	
	F. W. Flisher	C162, Vol. 52, No. 380, p. 150
	V. G. Sederman	2705, Vol. 41, No. 337, p. 206
5.13.	*To Construct a Triangle given A, and the Sums of Sides $c+a$, $b+a$* A. Lyngbye	3007, Vol. 46, No. 356, p. 135
5.14.	*Tangrams and Incommensurables*	
	C. Dudley Langford	1538, Vol. 25, No. 266, p. 233
5.15.	*From the Pentagon to the Dodecahedron*	
	F. C. Boon	1197, Vol. 20, No. 238, p. 153
	E. M. Bishop	C 97, Vol. 46, No. 358, p. 307
5.16.	*A Well-known Geometrical Problem*	
	J. St. C. L. Sinnadurai	3016, Vol. 46, No. 356, p. 143
	H. Lloyd	C185, Vol. 53, No. 383, p. 59
	M. G. Harman	C196, Vol. 53, No. 386, p. 398
5.17.	*When are Triangles Congruent?*	
	S. N. Collins	3261, Vol. 54, No. 388, p. 152

6 Trigonometry

6.1 Angles whose Sum is $\pi/4$

When considering the equilibrium of a uniform square lamina supported from one vertex and carrying a mass equal to that of the lamina at an adjacent vertex it is surprising to realise that the intriguing $\tan^{-1}(\frac{1}{2})+\tan^{-1}(\frac{1}{3})=\pi/4$ is geometrically obvious. In Fig. 1 with sides of lengths $4b$ and diagonals of length $4a$, it is clear that $\tan \angle OAG=\frac{1}{2}$, while $\tan \angle BAG=\frac{1}{3}$.

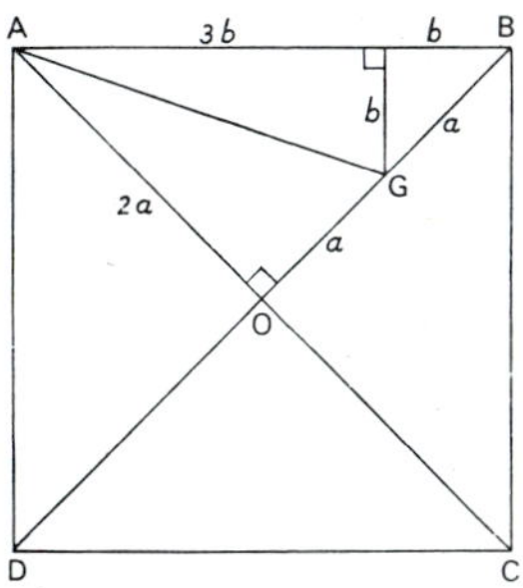

FIG. 1

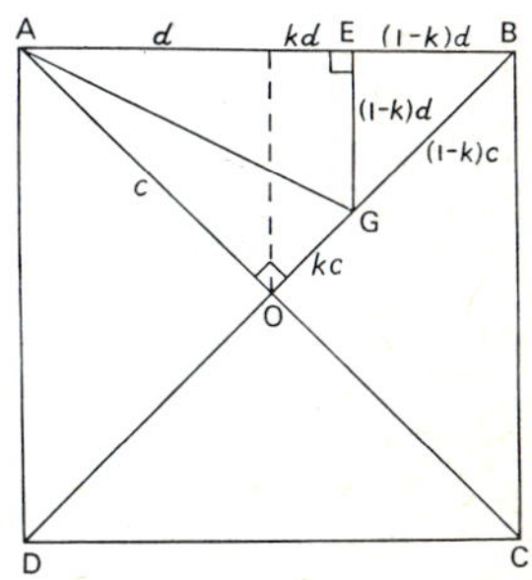

FIG. 2

It is easy to extend this result by taking G on OB so that $OG : OB = k : 1$ (Fig. 2).

Now $\tan \angle OAG = k$, while

$$\tan \angle EAG = \frac{GE}{AE} = \frac{1-k}{1+k}.$$

Hence
$$\tan^{-1} k + \tan^{-1}\left(\frac{1-k}{1+k}\right) = \frac{\pi}{4},$$

and we can confirm that k is not necessarily restricted to $0 < k < 1$.

The famous result $4\tan^{-1}\left(\frac{1}{5}\right)-\tan^{-1}\left(\frac{1}{239}\right)=\frac{\pi}{4}$ follows readily from the fact that

$$4\tan^{-1}\left(\frac{1}{5}\right) = 2\tan^{-1}\left(\frac{5}{12}\right) = \tan^{-1}\left(\frac{120}{119}\right).$$

Clearly if $k = 120/119$, we have the result in the form

$$4\tan^{-1}\left(\frac{1}{5}\right)+\tan^{-1}\left(\frac{1}{239}\right)=\frac{\pi}{4}.$$

Alternatively such results can easily be derived from a simple right-angled triangle:

FIG. 3

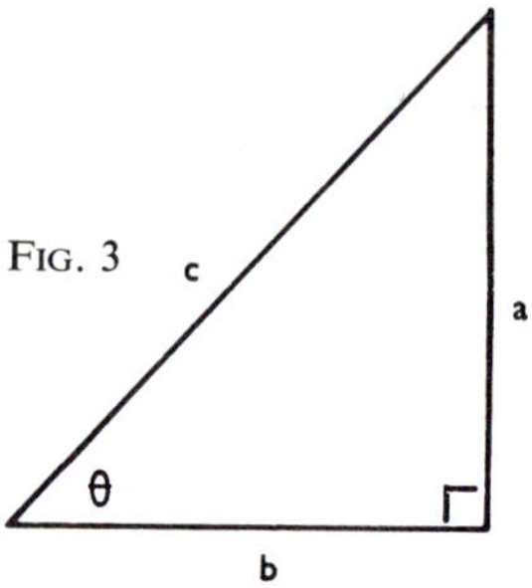

FIG. 4

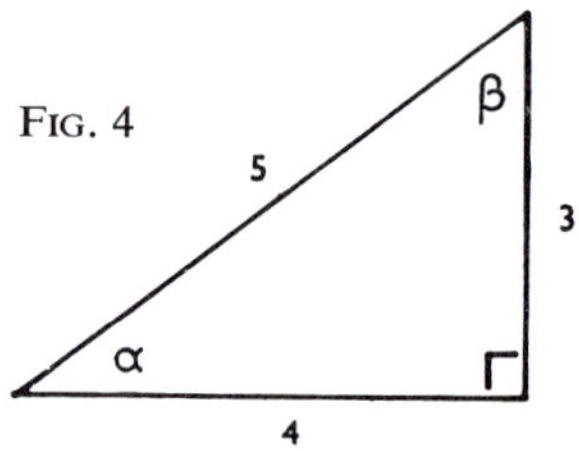

In any right-angled triangle (Fig. 3),

$$\tan\tfrac{1}{2}\theta = \frac{\sin\theta}{1+\cos\theta} = \frac{a}{c+b} = \frac{c-b}{a}.$$

Applying this to the 3, 4, 5 triangle (Fig. 4),

$$\tan\tfrac{1}{2}\alpha = \tfrac{1}{3} \quad \text{and} \quad \tan\tfrac{1}{2}\beta = \tfrac{1}{2}.$$

$$\therefore \tan^{-1}\tfrac{1}{2}+\tan^{-1}\tfrac{1}{3} = \tfrac{1}{4}\pi.$$

6.2 The Sine and Cosine of Obtuse Angles

Except for the very able pupil, the normal textbook approach to this topic is rather obscure. Most pupils have only thought of sines and cosines in connection with right-angled triangles (for usually this topic is dealt with just *before* the Sine and Cosine Rules). Even if the Sine and Cosine Rules for acute-angled triangles have been dealt with, the pupil still thinks of the sine and cosine of an angle in terms of the sides of a triangle in which the angle appears.

It is confusing to introduce all at once the concept of positive and negative lengths and immediately include a line whose length is always to be positive no matter what direction it has (even if it is almost parallel to a negative line); then to start working out the sine of an angle by considering an adjacent supplementary angle.

Since the definitions of sine and cosine of an obtuse angle (and definitions they are to most pupils at this stage!) are introduced at the same time as the Sine and Cosine Rules, why not use these Rules to justify our definitions?

In what follows, it is assumed that the Sine and Cosine Rules for acute-angled triangles only are known. The general case only is shown but numerical examples may be worked by the class.

The sine of an obtuse angle

Let us apply the Sine Rule to the triangle ABC (where $\angle CBA$ is

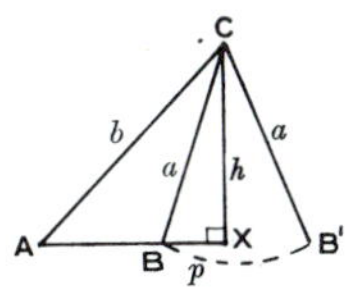

obtuse) and find what value $\sin \angle CBA$ gives.

If $$\frac{b}{\sin \angle CBA} = \frac{a}{\sin A}$$

then $$\sin \angle CBA = \frac{b \sin A}{a} = \frac{b}{a} \cdot \frac{h}{b} = \frac{h}{a} = \sin \angle CBX.$$

Hence if we take $\sin(180° - \theta)$ for $\sin \theta$ whenever θ is obtuse, the Sine Rule can be used for all possible triangles.

This means, however, that if we are given two sides, a and b (say), and the non-included angle A, and calculate B from $\sin B = \dfrac{b \sin A}{a}$, we must consider both the value obtained from tables and the supplement of this value as solutions.

But geometry supports this conclusion; for if we are given two sides and a non-included angle, we can sometimes construct two triangles (the ambiguous case) ACB and ACB' and

$$\angle ABC = 180° - \angle AB'C.$$

Even if there is only one triangle, i.e. if $a > b$, the obtuse solution of $\sin B = \dfrac{b \sin A}{a}$ is not absurd; it simply refers to another triangle having sides a and b, in which $\angle A$ is an exterior angle.

The cosine of an obtuse angle

As before, this is derived by finding what the Cosine Rule gives. Applying the formula to $\angle CBA$

$$\cos \angle CBA = \frac{a^2+c^2-b^2}{2ac},$$

but

$$b^2 = (c+p)^2+h^2$$
$$= c^2+2pc+a^2.$$

$$\therefore \cos \angle CBA = \frac{-2pc}{2ac} = -\frac{p}{a} = -\cos \angle CBX.$$

Hence if we take $-\cos(180°-\theta)$ for $\cos\theta$ whenever θ is obtuse, the Cosine Rule can be used for all possible triangles. There is no ambiguity here, of course, about the value of $\angle B$; $\angle B$ is acute, or obtuse, according as $\cos B$ is positive, or negative.

6.3 To Determine the Height of an Inaccessible Object

A number of textbooks include problems of the following type:

A man observes that the angle of elevation of the top of a tower is 20°; *he walks* 60 *ft directly towards the tower and sees that this angle is now* 30°. *Find the height of the tower*. (The figure is not to scale.)

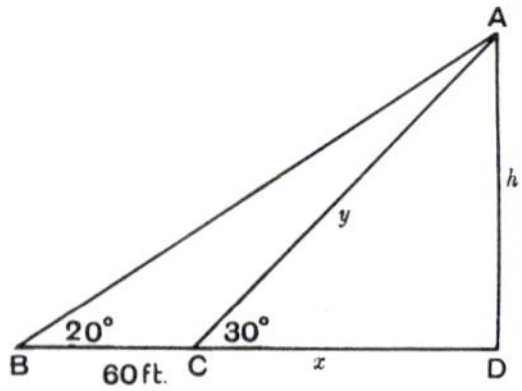

Many textbooks actually give an example like this as a worked example using the tangents of 20° and 30°, followed by the elimination of x by a method that many pupils find difficult.

Yet the practical users of trigonometry all did it as follows.

$$\angle BAC = 30° - 20° = 10°;$$

$$h/y = \sin 30°, \quad 60\,\text{ft}/\sin 10° = y/\sin 20°.$$

Hence
$$h = \frac{\sin 30° \sin 20°}{\sin 10°} \,.\, 60\,\text{ft}.$$

Now, how can this anomaly be explained? Presumably the authors either want to show pupils as soon as possible that trigonometry can do all sorts of things they had not thought possible, or they are trying to find other examples to make the pupil familiar with the use of the tangent. But does either reason justify asking a pupil to do a problem by a method that would not be used by the practical man, particularly when it is sufficiently awkward to put quite a number of pupils off in despair? Surely it would be better to give more simple examples for practice and pass on as rapidly as possible to the Sine Rule when problems of this kind can then come in.

6.4 The Cosine Rule

(i) In any triangle ABC,

$$a = b\cos C + c\cos B,$$

with similar formulae for b and c.

$$\begin{aligned} b^2+c^2-2bc\cos A &= b(b-c\cos A)+c(c-b\cos A) \\ &= ba\cos C + ca\cos B \\ &= a(b\cos C + c\cos B) \\ &= a^2. \end{aligned}$$

The following variant has the advantage of starting on the a^2 side.

$$\begin{aligned} a^2 &= a(b\cos C + c\cos B) \\ &= ba\cos C + ca\cos B \\ &= b(b-c\cos A)+c(c-b\cos A)\text{ etc.} \end{aligned}$$

(ii) But at this stage in the course the pupils still really need a figure including an altitude from B or C. In *any* figure containing the altitude from B, Pythagoras' Theorem gives

$$a^2 = (b-c\cos A)^2 + (c\sin A)^2 \text{ etc.}$$

This seems preferable to the juggling with p (altitude) and x (projection) that often goes on.

(iii) It also suggests an amusing refinement, in which we dispense with the figure.

$$\begin{aligned} a^2 &= a^2(\cos^2 C+\sin^2 C) \\ &= (a\cos C)^2+(a\sin C)^2 \\ &= (b-c\cos A)^2+(c\sin A)^2, \text{etc.} \end{aligned}$$

Problems which involve Latitude, Longitude and Great Circles often prove difficult for pupils, but the following two ideas should prove helpful.

6.5 Model to Illustrate Latitude and Longitude†

Materials required: thin cardboard and 'Sellotape'.

Any convenient measurements, provided $e^2 = a^2-d^2$. For example (in cm) $a = 10$, $b = 1{\cdot}5$, $c = 5$, $d = 6$, $e = 8$; α about 30°, β, γ about 50°. (See opposite).

Cut along continuous lines.

For ease of movement, and to allow for thickness of cardboard, narrow strips should be cut away along the internal cutting lines in II, IV, V; radii in III should be slightly less than b, e, greater than c (e.g. 1·4, 5·1, 7·9 or even 1·3, 5·2 if constructors are not very accurate! It is better to cut away too much than too little).

1. Bend flaps x, of II, alternately backwards and forwards along AB, and stick to I, matching A, O, B; planes I and II perpendicular.
2. Insert III, matching X, Y, P; planes I and III parallel.
3. Bend IV backwards and forwards along PN.

   Insert IV (bending when necessary) with U and W on opposite sides of II (AON), making two (rotatable) planes perpendicular to I.

   (Alternatively, construction can be made easier by using two pieces V in place of IV.)
4. Fix III to II (at Y say).

   IV may be attached to II along PO and the two parts of II at N joined through a small cut in IV.

† Another model can be found in: *Schools Mathematics Project*, 5, Cambridge University Press, p. 34.

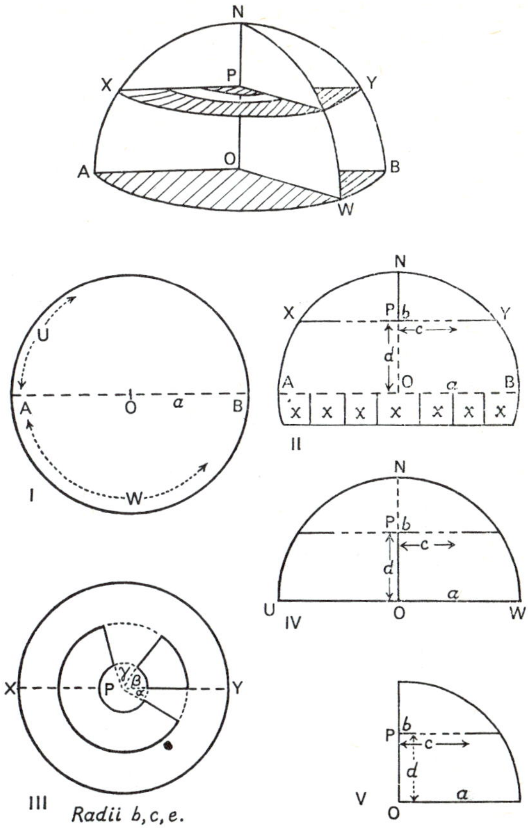

Note: Any lines (such as *OX*) required on finished model should be ruled on pieces before assembly.

Any number of loose pieces V may be made and used to draw on in particular problems.

The whole model folds flat with *U* and *W* at *A*.

N.B. The angle of latitude should be indicated on II, IV and V before the model is assembled.

6.6 Great Circles on a Sphere †

In this age of air travel, most secondary school courses include some study of the geometry of the sphere, with such topics as latitude, longitude and great circles. The distance along a circle of latitude and along a great circle joining two points on the same circle of latitude are often calculated or found by drawing. The general case of the great circle distance between any two points of known latitude and longitude is often omitted, but the diagram below shows that it is scarcely more difficult.

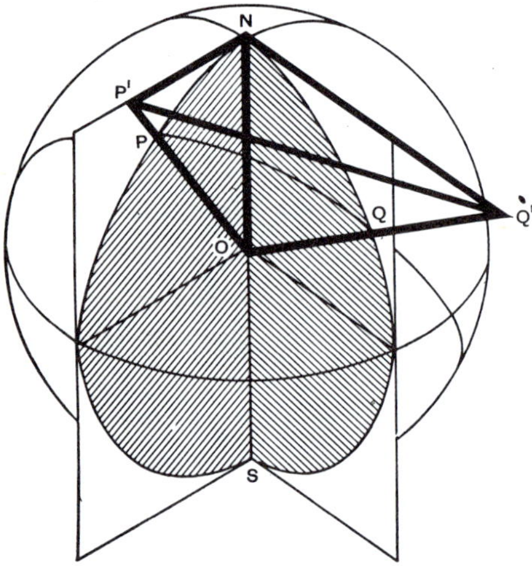

Consider the intersection of the plane of the required great circle with the plane touching the sphere at the north (or south) pole. For ease of description we shall call the latter plane horizontal and the axis of the earth vertical. P and Q are the two points whose latitudes and longitude difference are known. We draw (a) the vertical section through N and P: (b) the vertical section through N and Q. These give us NP' and NQ'. We then draw (c) the horizontal triangle $P'NQ'$, taking $\angle P'NQ'$ as the given difference in longitude. Finally we draw (d) the section through P', Q', P, Q, and O; this contains the great circle through P and Q. To do this we get OP' and OQ' from (a) and (b) and $P'Q'$ from (c).

† Further references to great circles, Moakes. A. J., Croome. P. D., Phillips, T. C., *Pattern and Power of Mathematics*, 6, Macmillan, p. 240.

The number of minutes in the angle POQ is the great circle distance in nautical miles from P to Q. We can also find from the same figures the most northerly point of the track, or the latitude and longitudes of points dividing the track in any given ratio, and so on. Should PQ cross the equator, we first find the great circle between P and the point antipodal to Q.

6.7 A Model to Demonstrate the Variation of sin x and cos x†

The model consists of two plywood circles radii 10 cm and 20 cm, the smaller passing through the centre O of the larger. If OP be a diameter of the smaller circle and AOB', BOB' two perpendicular diameters of the larger graduated from -1 to $+1$, the intersections of the smaller circle with AOA', BOB' of course record $\cos x$ and $\sin x$ when x is the angle POA.

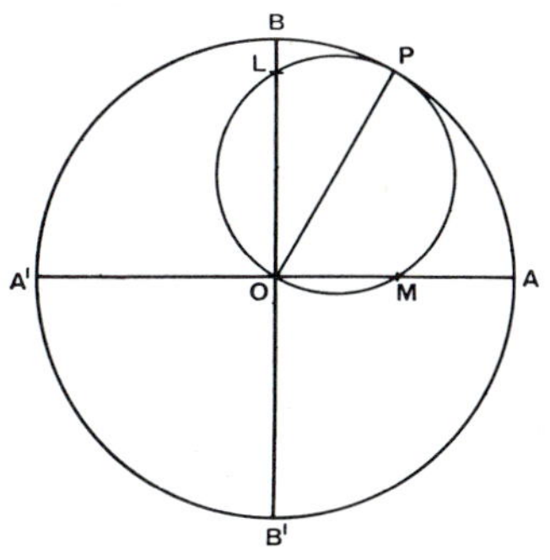

A quite cheap thin plywood can be used, and prepared for drawing with Indian ink by painting in white undercoating very much thinned with turpentine. Incidentally this, well glass-papered, is a useful material for drawings which are to have a more or less permanent existence in the classroom.

An effective O.H.P. transparency can be made, based on this idea. With suitably marked axes, sines and cosines can be determined to an accuracy of 0·01.

No chapter on trigonometry would be complete without some reference to the addition formulae.

† *Also* Joint, W. Howard, Note 1766, *Mathematical Gazette*, Vol. 28, No. 282, p. 194.

6.8 The Addition Formulae

The first four notes are based on triangle properties.
$\sin(\alpha+\beta)$ from Sine Rule; $\cos(\alpha+\beta)$ from Cosine Rule

I $\sin(\alpha+\beta)$.

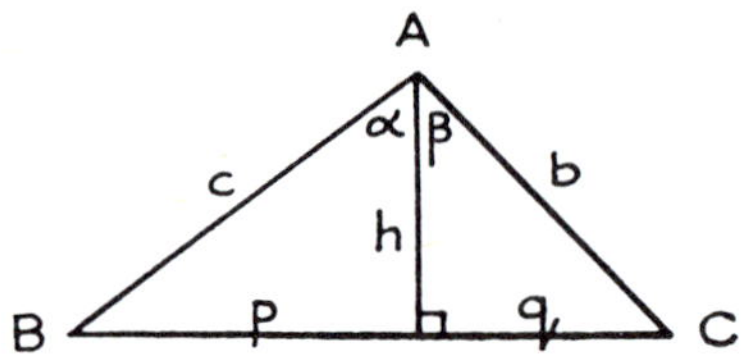

In the notation of the diagram, the Sine Rule for triangle *ABC* gives

$$\frac{p+q}{\sin(\alpha+\beta)} = \frac{c}{\cos\beta} = \frac{b}{\cos\alpha}$$

$$= \frac{1}{\cos\beta}\cdot\frac{p}{\sin\alpha} = \frac{1}{\cos\alpha}\cdot\frac{q}{\sin\beta}$$

$$= \frac{p+q}{\sin\alpha\cos\beta+\cos\alpha\sin\beta}; \text{ etc.}$$

II $\cos(\alpha+\beta)$

$$a^2 = b^2+c^2-2bc\cos A$$

$$\therefore\ (p+q)^2 = b^2+c^2-2bc\cos(\alpha+\beta).$$

$$\therefore\ 2bc\cos(\alpha+\beta) = (b^2-q^2)+(c^2-p^2)-2pq$$

$$= h^2+h^2-2pq.$$

$$\therefore\ \cos(\alpha+\beta) = \frac{h}{c}\cdot\frac{h}{b}-\frac{p}{c}\cdot\frac{q}{b}$$

$$= \cos\alpha\cos\beta-\sin\alpha\sin\beta.$$

Notes:

(i) The proofs hold for all acute angles α, β.

(ii) With angle *ACB* obtuse, the formulae for $\sin(\alpha-\beta)$ and $\cos(\alpha-\beta)$ are obtained similarly. Again α, β must be acute, and consequently $\alpha-\beta$ also.

(iii) No. **I** is virtually the first proof given on p. 63 of the Association's *Trigonometry Report*,† where explicit use is made of the formula $a = b\cos C + c\cos B$ (which happens not to be included in the syllabus for the Scottish Certificate of Education!). The proof of No. **II** is a little simpler if this formula is not used.

III The following proofs depend on the easily demonstrated formulae $\Delta = \frac{1}{2}ab\sin C$ and the associated

$$a/\sin A = b/\sin B = c/\sin C = 2\Delta/abc.$$

Many books use the former for proving the formulae for $\sin(A \pm B)$ but a totally different method for $\cos(A \pm B)$.

1. To prove that

$$\cos \overline{A+B} = \cos A \cos B - \sin A \sin B.$$

With the diagram for proving the formula for $\sin \overline{A+B}$ draw NP perpendicular to LM.

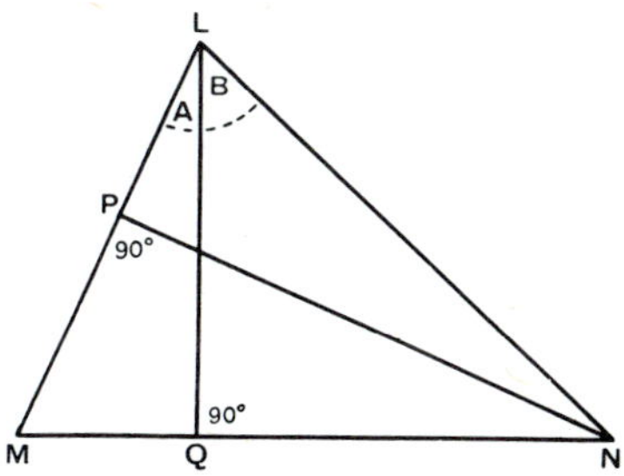

We have

$$\triangle PNL = \triangle MNL - \triangle MNP,$$

or $$\tfrac{1}{2}PN \,.\, LN\sin\{\tfrac{1}{2}\pi - (A+B)\} = \tfrac{1}{2}MN \,.\, LN\sin(\tfrac{1}{2}\pi - B) - \tfrac{1}{2}MN \,.\, PN\sin A;$$

that is,

$$\cos(A+B) = (MN/PN)\cos B - (MN/LN)\sin A$$

$$= \frac{\cos B}{\cos A} - \frac{\sin(A+B)}{\cos A}\sin A$$

† Mathematical Association, *The Teaching of Trigonometry in Schools*, Bell, p. 63.

$$= (\cos B - \sin^2 A \cos B - \cos A \sin A \sin B)/\cos A$$

$$= (\cos B \cos^2 A - \cos A \sin A \sin B)/\cos A$$

$$= \cos A \cos B - \sin A \sin B.$$

2. To prove that

$$\cos(A-B) = \cos A \cos B + \sin A \sin B.$$

In the above diagram let $\angle MLN = A$, $\angle MLQ = B$, so that $\angle QLN = A-B$. Then

$$\triangle LNM = \triangle PNL + \triangle PNM.$$

Thus

$$\tfrac{1}{2}MN\,.\,LN\,.\,\sin\{\tfrac{1}{2}\pi-(A-B)\} = \tfrac{1}{2}PN\,.\,LN\sin(\tfrac{1}{2}\pi - A) + \tfrac{1}{2}MN\,.\,PN\sin B;$$

that is,

$$\cos(A-B) = (PN/MN)\cos A + (PN/LN)\sin B$$
$$= \cos A \cos B + \sin A \sin B.$$

IV Here are proofs based only on the area of a triangle, and are the result of a remark made by a pupil.

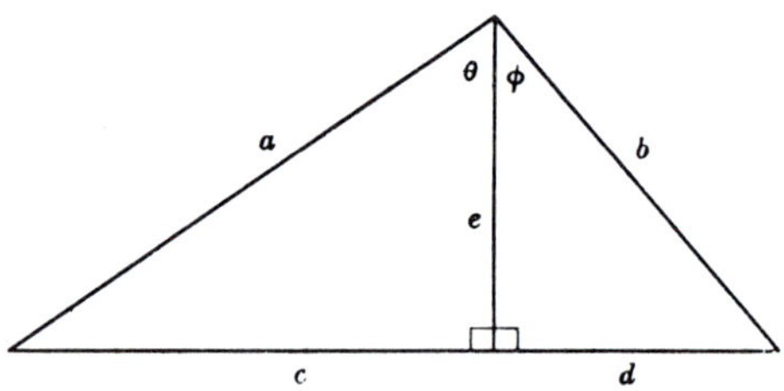

Writing down the area of the whole figure in two ways, we have

$$\tfrac{1}{2}ab\sin(\theta+\phi) = \tfrac{1}{2}ce + \tfrac{1}{2}ed.$$

Hence $$\sin(\theta+\phi) = ce/ab + ed/ab$$

$$= \sin\theta\cos\phi + \cos\theta\sin\phi.$$

The modification for $\sin(\theta-\phi)$ is obvious: fold the figure about the line e and consider the area of the triangle which is then the difference of two right-angled triangles.

For the cosine, make a small change in the lettering of the figure, replacing θ by $\frac{1}{2}\pi-\theta$. Then as before

$$\tfrac{1}{2}ab\sin(\phi+\tfrac{1}{2}\pi-\theta)=\tfrac{1}{2}ce+\tfrac{1}{2}ed.$$

Hence
$$\cos(\theta-\phi) = ce/ab+ed/ab$$
$$= \cos\theta\cos\phi+\sin\theta\sin\phi,$$

and the same modification as above will give us the formula for $\cos(\theta+\phi)$.

The following is a more elegant and shortened form of the above proof.

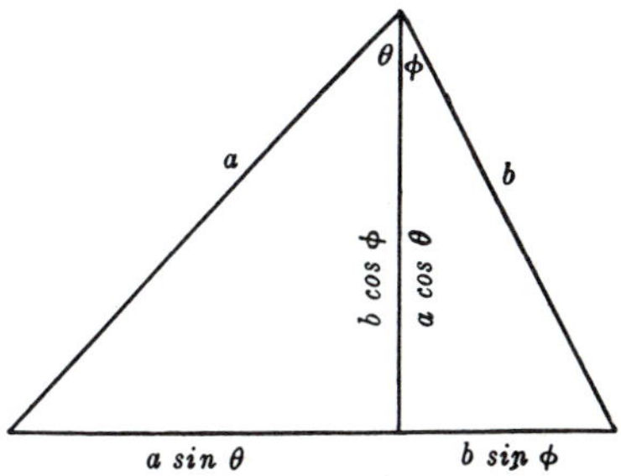

$$\tfrac{1}{2}ab\sin(\theta+\phi) = \triangle = \tfrac{1}{2}a\sin\theta \,.\, b\cos\phi+\tfrac{1}{2}a\cos\theta \,.\, b\sin\phi.$$

V It is possible to generalise the results that hold for acute angles.

The circular functions for the general angle can be dealt with very simply by defining them as such that the functional relations obtained for acute angles will still hold in general. This is a more logical procedure than the formation of intricate definitions to be followed by more complicated proofs of the addition formulae.

The essential relations for sine and cosine are

$$\sin A = \cos(90°-A), \qquad (1)$$

$$\sin(A+B) = \sin A \,.\, \cos B+\cos A \,.\, \sin B. \qquad (2)$$

If we accept the values obtained by the usual methods for angles from 0° to 90° inclusive and let (1) and (2) hold for *all angles*, then the argument may be as follows.

In (1) put

$$A = 90°-B \text{ and get } \sin(90°-B) = \cos A, \qquad (a)$$

$$\sin(90°+A) = \sin 90° \,.\, \cos A + \cos 90° \,.\, \sin A = \cos A, \qquad \text{(b)}$$

$$\cos(-A) = \sin(90°-(-A)) = \sin(90°+A) = \cos A, \qquad \text{(c)}$$

$$0 = \sin 0° = \sin(A+(-A)) = \sin A \,.\, \cos(-A) + \cos A \,.\, \sin(-A)$$
$$= \cos A(\sin A + \sin(-A)),$$
$$\therefore\ \sin A = -\sin(-A), \qquad \text{(d)}$$

$$\sin(A-B) = \sin(A+(-B)) = \sin A \,.\, \cos(-B) + \cos A \,.\, \sin(-B)$$
$$= \sin A \,.\, \cos B - \cos A \,.\, \sin B, \qquad \text{(e)}$$

$$\cos(A+B) = \sin(\overline{90°-A}-B) = \sin(90°-A) \,.\, \cos B - \cos(90°-A) \,.\, \sin B$$
$$= \cos A \,.\, \cos B - \sin A \,.\, \sin B, \qquad \text{(f)}$$

$$\cos(A-B) = \cos A \,.\, \cos(-B) - \sin A \,.\, \sin(-B)$$
$$= \cos A \,.\, \cos B + \sin A \,.\, \sin B. \qquad \text{(g)}$$

Also $1 = \cos 0° = \cos(A-A) = \cos^2 A + \sin^2 A.$ (h)

Thus, the addition and Pythagorean formulae, and all derived formulae, for all the circular functions hold generally.

The Periodic Nature is established in the following sequence:

$$\sin A = \cos(90°-A) = \sin(90° + \overline{90°-A}) = \sin(180°-A),$$

$$\sin(180°+A) = \sin(90° + \overline{90°+A}) = \cos(90°+A) = -\sin A,$$

$$\sin(360°+A) = \sin(180° + \overline{180°+A}) = -\sin(180°+A) = \sin A,$$

whence, if $360°+A = B$ then $\sin B = \sin(B-360°)$,

$$\cos A = \sin(90°+A) = \sin(360° + \overline{90°+A}) = \sin(90° + \overline{360°+A})$$
$$= \cos(360°+A).$$

VI A general proof based on the scalar product of vectors†

With reference to axes Ox and Oy as shown, (in the fig. opposite) let the position vectors of A and B be $\vec{OA} = \begin{pmatrix} x_1 \\ y_1 \end{pmatrix}$ and $\vec{OB} = \begin{pmatrix} x_2 \\ y_2 \end{pmatrix}$.

Then $$\vec{OA} \,.\, \vec{OB} = OA \,.\, OB \,.\, \cos(\theta_1-\theta_2),$$

† *Also* Mathematical Association, *The Teaching of Trigonometry in Schools*, Bell, p. 67. A more explicit proof is given in *Schools Mathematics Project*, Additional Mathematics, 2, Cambridge University Press, p. 306.

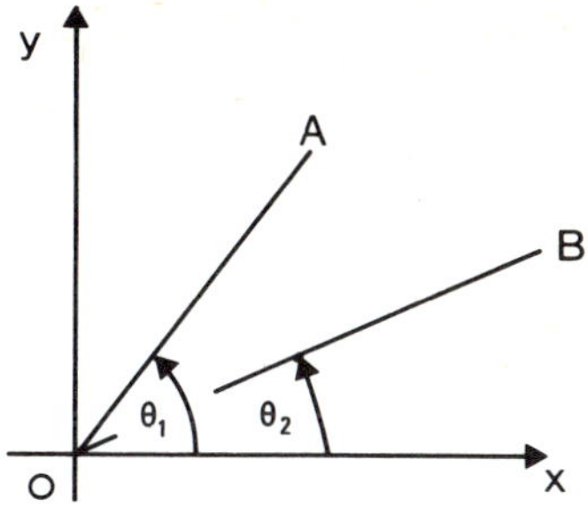

but $\overrightarrow{OA} . \overrightarrow{OB} = x_1 . x_2 + y_1 . y_2$

$$= OA \cos\theta_1 . OB \cos\theta_2 + OA \sin\theta_1 . OB \sin\theta_2,$$

hence $\qquad \cos(\theta_1 - \theta_2) = \cos\theta_1 . \cos\theta_2 + \sin\theta_1 . \sin\theta_2.$

The other results then follow.

A proof involving the rotation of vectors using matrices can be found elsewhere.‡

VII The following proof of the addition theorems is general and is easier to explain and easier for students to appreciate than the usual proof involving vectors. No reference to the method has been found in any modern textbook on elementary trigonometry.

It is most natural to begin by establishing the formula for $\sin(A-B)$.

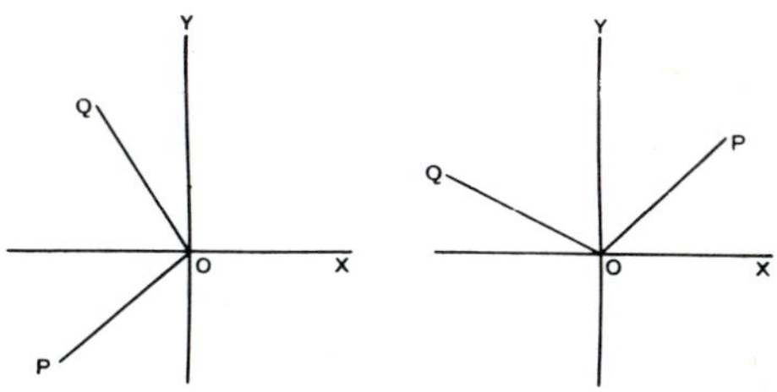

Let A and B be angles of any magnitude and let OP and OQ be unit vectors such that $\angle XOP$ and $\angle XOQ$ are coterminal with A and B respectively. Then the point P is $(\cos A, \sin A)$, the point Q is $(\cos B, \sin B)$, $\angle QOP$ is coterminal with $A-B$, and the area of the

‡ *Schools Mathematics Project*, Additional Mathematics, 2, Cambridge University Press, p. 317.

triangle OQP is $\frac{1}{2}\sin(A-B)$ in all cases, the area being regarded as positive if the triangle is described in a counter-clockwise sense and otherwise negative.

The area of a triangle with vertices $(0, 0)$ (x_1, y_1) (x_2, y_2) is

$$\frac{1}{2}\begin{vmatrix} x_1, & y_1 \\ x_2, & y_2 \end{vmatrix},$$

the same convention as to sign being adopted as above. (The determinant notation is not essential.)

Thus

$$\begin{aligned} 2\triangle OQP &= \sin(A-B) \\ &= \begin{vmatrix} \cos B, & \sin B \\ \cos A, & \sin A \end{vmatrix} \\ &= \sin A\cos B - \cos A\sin B. \end{aligned}$$

The formulae for $\sin(A+B)$ and $\cos(A\pm B)$ follow in the usual way.

REFERENCES

6.1. *Angle whose Sum is $\pi/4$*
A. J. G. May — C187, Vol. 53, No. 384, p. 157
W. R. S. North — C231, Vol. 54, No. 390, p. 377

6.2. *The Sine and Cosine of Obtuse Angles*
M. A. Stamp — C 90, Vol. 46, No. 356, p. 154

6.3. *To Determine the Height of an Inaccessible Object*
C. D. Langford — 1717, Vol. 28, No. 279, p. 69

6.4. *The Cosine Rule*
P. Gant — 2556, Vol. 39, No. 329, p. 239
A. G. Sillitto — C 9, Vol. 42, No. 339, p. 64

6.5. *Model to Illustrate Latitude and Longitude*
Jane Birkett — C174, Vol. 52, No. 382, p. 372

6.6. *Great Circles on a Sphere*
R. C. Lyness — 2882, Vol. 44, No. 348, p. 115

6.7. *Model to Demonstrate the Variation of sin x and cos x*
A. G. Granston Richards — 1407, Vol. 23, No. 256, p. 391

6.8. *The Addition Formulae*
A. G. Sillitto — C101, Vol. 47, No. 359, p. 58
J. Pedoe — 1409, Vol. 23, No. 256, p. 392
A. R. Pargeter — 1680, Vol. 27, No. 276, p. 185
G. A. Clarkson & C. E. Kemp — 1747, Vol. 28, No. 282, p. 181
A. Robson — 1438, Vol. 24, No. 258, p. 56
F. Jackson — 2168, Vol. 34, No. 310, p. 286
R. Smart — 1450, Vol. 24, No. 259, p. 116

7 Vectors

7.1 The Use of Vectors for Finding OH^2, OI^2 in a Triangle

(i) OH^2

For a triangle ABC, and an origin at the circumcentre, taking α, β, γ to be the lengths of the sides,

$$a^2 = b^2 = c^2 = R^2 \quad \text{where } R \text{ is the circumradius.}$$

For centroid G, $\mathbf{g} = \frac{1}{3}(\mathbf{a}+\mathbf{b}+\mathbf{c})$.
Consider H such that $\mathbf{h} = \mathbf{a}+\mathbf{b}+\mathbf{c} = 3\mathbf{g}$ $(\mathbf{OH} = 3\mathbf{OG})$
Then $(\mathbf{h}-\mathbf{a}) \cdot (\mathbf{b}-\mathbf{c}) = (\mathbf{b}+\mathbf{c}) \cdot (\mathbf{b}-\mathbf{c}) = b^2 - c^2 = 0$

$\therefore$ HA is perpendicular to BC etc.

$\therefore$ H is the orthocentre.

$$OH^2 = h^2 = (\mathbf{a}+\mathbf{b}+\mathbf{c}) \cdot (\mathbf{a}+\mathbf{b}+\mathbf{c}) = \Sigma a^2 + 2\Sigma(\mathbf{b} \cdot \mathbf{c})$$

i.e. $OH^2 = 3R^2 + 2R^2 \Sigma \cos 2A = R^2(1 - 8\cos A \cos B \cos C)$
(since $\cos^2 A + \cos^2 B + \cos^2 C + 2\cos A \cos B \cos C = 1$).

(ii) OI^2

The incentre I is the centre of gravity of weights α, β, γ at A, B, C.

$$\mathbf{i} = \frac{\alpha\mathbf{a}+\beta\mathbf{b}+\gamma\mathbf{c}}{2s} \quad \text{where} \quad 2s = \alpha+\beta+\gamma$$

$$\begin{aligned}
\therefore\ 4s^2 i^2 &= (\alpha\mathbf{a}+\beta\mathbf{b}+\gamma\mathbf{c}) \cdot (\alpha\mathbf{a}+\beta\mathbf{b}+\gamma\mathbf{c}) \\
&= \Sigma\alpha^2 a^2 + 2\Sigma\beta\gamma\mathbf{b} \cdot \mathbf{c} = R^2(\Sigma\alpha^2 + 2\Sigma\beta\gamma\cos 2A) \\
&= R^2(\{\Sigma\alpha\}^2 - 4\Sigma\beta\gamma\sin^2 A) \\
&= R^2\left(4s^2 - 4\Delta\frac{(\alpha+\beta+\gamma)}{R}\right) \quad (\Delta \text{ is area of triangle}) \\
&= 4s^2\left(R^2 - \frac{2R\Delta}{s}\right)
\end{aligned}$$

$$\therefore\ OI^2 = R^2 - 2Rr \quad (r \text{ is inradius})$$

7.2 Triple Vector Product Formula

I Since $[(\mathbf{a}\,.\,\mathbf{b})\mathbf{b}-(\mathbf{b}\,.\,\mathbf{b})\mathbf{a}]\,.\,\mathbf{b}$

$$= (\mathbf{a}\,.\,\mathbf{b})(\mathbf{b}\,.\,\mathbf{b})-(\mathbf{b}\,.\,\mathbf{b})(\mathbf{a}\,.\,\mathbf{b}) = 0$$

$[(\mathbf{a}\,.\,\mathbf{b})\mathbf{b}-(\mathbf{b}\,.\,\mathbf{b})\mathbf{a}]$ is in the plane of **a** and **b** and is perpendicular to **b**

$$\therefore\ (\mathbf{a} \wedge \mathbf{b}) \wedge \mathbf{b} = \lambda[(\mathbf{a}\,.\,\mathbf{b})\mathbf{b}-(\mathbf{b}\,.\,\mathbf{b})\mathbf{a}]$$

$\therefore$ scalar multiplying by **a**,

$$[(\mathbf{a} \wedge \mathbf{b}) \wedge \mathbf{b}]\,.\,\mathbf{a} = \lambda[(\mathbf{a}\,.\,\mathbf{b})\mathbf{b}-(\mathbf{b}\,.\,\mathbf{b})\mathbf{a}]\,.\,\mathbf{a}$$

i.e. $$(\mathbf{b} \wedge \mathbf{a})\,.\,(\mathbf{a} \wedge \mathbf{b}) = \lambda[(\mathbf{a}\,.\,\mathbf{b})^2-(\mathbf{b}\,.\,\mathbf{b})(\mathbf{a}\,.\,\mathbf{a})]$$

i.e. $$-(\mathbf{a} \wedge \mathbf{b})\,.\,(\mathbf{a} \wedge \mathbf{b}) = \lambda[(\mathbf{a}\,.\,\mathbf{b})^2-b^2a^2].$$

Now $|(\mathbf{a} \wedge \mathbf{b})\,.\,(\mathbf{a} \wedge \mathbf{b})| = a^2b^2\sin^2\theta$ where θ is the angle between **a** and **b** and

$$|\lambda\,[(\mathbf{a}\,.\,\mathbf{b})^2-b^2a^2]| = \lambda(a^2b^2-a^2b^2\cos^2\theta) = \lambda(a^2b^2\sin^2\theta).$$

Hence $\lambda = 1$

Now, let $\mathbf{c} = \lambda_1\mathbf{a}+\lambda_2\mathbf{b}+\lambda_3(\mathbf{a} \wedge \mathbf{b})$

then $\mathbf{a}\,.\,\mathbf{c} = \lambda_1 a^2+\lambda_2(\mathbf{a}\,.\,\mathbf{b})$

and $\mathbf{b}\,.\,\mathbf{c} = \lambda_1(\mathbf{a}\,.\,\mathbf{b})+\lambda_2 b^2.$

Thus

$$\begin{aligned}(\mathbf{a} \wedge \mathbf{b}) \wedge \mathbf{c} &= \lambda_1(\mathbf{a} \wedge \mathbf{b}) \wedge \mathbf{a}+\lambda_2(\mathbf{a} \wedge \mathbf{b}) \wedge \mathbf{b}\\ &= \lambda_1[(\mathbf{a}\,.\,\mathbf{a})\mathbf{b}-(\mathbf{a}\,.\,\mathbf{b})\mathbf{a}]+\lambda_2[(\mathbf{a}\,.\,\mathbf{b})\mathbf{b}-(\mathbf{b}\,.\,\mathbf{b})\mathbf{a}]\end{aligned}$$

$$\therefore\ (\mathbf{a} \wedge \mathbf{b}) \wedge \mathbf{c} = (\mathbf{a}\,.\,\mathbf{c})\mathbf{b}-(\mathbf{b}\,.\,\mathbf{c})\mathbf{a}.\dagger$$

Alternatively, the result can be proved using the following lemma.

Lemma

$$(\mathbf{a} \wedge \mathbf{b}) \wedge \mathbf{b} = (\mathbf{a}\,.\,\mathbf{b})\mathbf{b}-b^2\mathbf{a}$$

$\mathbf{d} = (\mathbf{a} \wedge \mathbf{b}) \wedge \mathbf{b}$ is perpendicular to **b** and in the plane containing **a**

† For a similar treatment see Spain, B., *Vector Analysis*, Van Nostrand, 1965, p. 18.

and **b**, the angle between the positive directions of **d** and **a** being obtuse.

Hence the following vector diagram holds

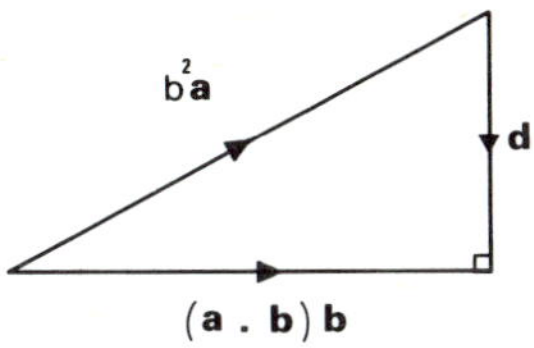

thus proving the lemma.

Now $\quad (\mathbf{a} \wedge \mathbf{b}) \wedge \mathbf{c} = \lambda[(\mathbf{a} \,.\, \mathbf{c})\mathbf{b} - (\mathbf{b} \,.\, \mathbf{c})\mathbf{a}]$

or
$$\begin{aligned}[(\mathbf{a} \,.\, \mathbf{c})\mathbf{b} - (\mathbf{b} \,.\, \mathbf{c})\mathbf{a}] \,.\, \mathbf{b} &= [(\mathbf{a} \wedge \mathbf{b}) \wedge \mathbf{c}] \,.\, \mathbf{b} \\ &= [\mathbf{b} \wedge (\mathbf{a} \wedge \mathbf{b})] \,.\, \mathbf{c} \\ &= [b^2\mathbf{a} - (\mathbf{a} \,.\, \mathbf{b})\mathbf{b}] \,.\, \mathbf{c},\end{aligned}$$

whence $\lambda = 1$

$$\therefore (\mathbf{a} \wedge \mathbf{b}) \wedge \mathbf{c} = (\mathbf{a} \,.\, \mathbf{c})\mathbf{b} - (\mathbf{b} \,.\, \mathbf{c})\mathbf{a}.$$

II The following is a proof of the triple vector product formula involving only elementary geometry, and a minimum of calculation.

$(\mathbf{a} \wedge \mathbf{b}) \wedge \mathbf{c}$ is in the plane of **a** and **b** and perpendicular to **c**,

$$\therefore (\mathbf{a} \wedge \mathbf{b}) \wedge \mathbf{c} = \lambda[(\mathbf{a} \,.\, \mathbf{c})\mathbf{b} - (\mathbf{b} \,.\, \mathbf{c})\mathbf{a}], \tag{1}$$

where λ is a scalar, which must be shown to be a constant and equal to 1.

Both $(\mathbf{a} \wedge \mathbf{b}) \wedge \mathbf{c}$ and $[(\mathbf{a} \,.\, \mathbf{c})\mathbf{b} - (\mathbf{b} \,.\, \mathbf{c})\mathbf{a}]$ are linear and homogeneous in all three vectors.

If **c** is expressed as the sum of its components perpendicular to the plane of **a** and **b** and in the plane, the component perpendicular to the plane contributes nothing to either side of (1). Hence **c** may be replaced by its component in the plane of **a** and **b** without affecting λ. Similarly the component of **a** parallel to **b** can be ignored, leaving the component of **a** perpendicular to **b**. Both **a** and **b** can be replaced by unit vectors in their respective directions without altering λ.

So it is sufficient to prove $\lambda = 1$ in the special case where **a**, **b** and **c** are coplanar and **a** and **b** are orthogonal unit vectors. In this case $(\mathbf{a} \wedge \mathbf{b})$ is a unit vector perpendicular to the plane of **a** and **b** such that

$\{\mathbf{a}, \mathbf{b}, (\mathbf{a} \wedge \mathbf{b})\}$ form a right-handed system. $[(\mathbf{a} \wedge \mathbf{b}) \wedge \mathbf{c}]$ is in the plane of **a**, **b**, **c** such that $\{(\mathbf{a} \wedge \mathbf{b}), \mathbf{c}, [(\mathbf{a} \wedge \mathbf{b}) \wedge \mathbf{c}]\}$ form a right-handed system, and is obtained from **c** by rotating through 90° in the plane of **a**, **b**, **c** in a direction the same as that from **a** to **b**.

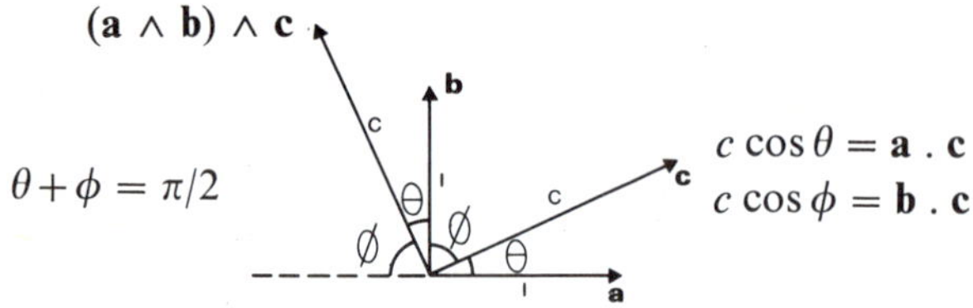

From the diagram,

$$(\mathbf{a} \wedge \mathbf{b}) \wedge \mathbf{c} = c\cos\theta\,\mathbf{b} - c\cos\phi\,\mathbf{a} = (\mathbf{a}\,.\,\mathbf{c})\mathbf{b} - (\mathbf{b}\,.\,\mathbf{c})\mathbf{a}$$

III Below is one of the simplest proofs of the triple vector product expansion.

Let $(\mathbf{a} \wedge \mathbf{b}) \wedge \mathbf{c} = X\mathbf{a} + Y\mathbf{b}$, since the triple product is in the plane of **a** and **b**. Suppose **f** is in the plane of **a** and **b** and is perpendicular to **b**.

Then
$$X(\mathbf{a}\,.\,\mathbf{f}) = [(\mathbf{a} \wedge \mathbf{b}) \wedge \mathbf{c}]\,.\,\mathbf{f}$$
$$= \mathbf{c}\,.\,[\mathbf{f} \wedge (\mathbf{a} \wedge \mathbf{b})].$$

If θ is the angle between **a** and **b**, $\mathbf{f} \wedge (\mathbf{a} \wedge \mathbf{b})$ has magnitude $abf\sin\theta$, and direction along $-\mathbf{b}$.

Hence, $\mathbf{f} \wedge (\mathbf{a} \wedge \mathbf{b}) = -(\mathbf{a} \wedge \mathbf{f})\,.\,\mathbf{b}$ (if ϕ is the angle between **a** and **f**, $\sin\theta = \cos\phi$)

$$\therefore\ X(\mathbf{a}\,.\,\mathbf{f}) = -(\mathbf{c}\,.\,\mathbf{b})(\mathbf{a}\,.\,\mathbf{f})$$
$$\therefore\ X = -\mathbf{c}\,.\,\mathbf{b}.$$

Similarly $Y = \mathbf{a}\,.\,\mathbf{c}$

hence $(\mathbf{a} \wedge \mathbf{b}) \wedge \mathbf{c} = (\mathbf{a}\,.\,\mathbf{c})\mathbf{b} - (\mathbf{b}\,.\,\mathbf{c})\mathbf{a}.$

IV The following proves the result for unit vectors—it can easily be extended by the introduction of scalars.

(i) special case when $\mathbf{b} = \mathbf{c}$

$(\mathbf{a} \wedge \mathbf{c}) \wedge \mathbf{c}$ is in the plane of **a** and **c** and perpendicular to **c**.

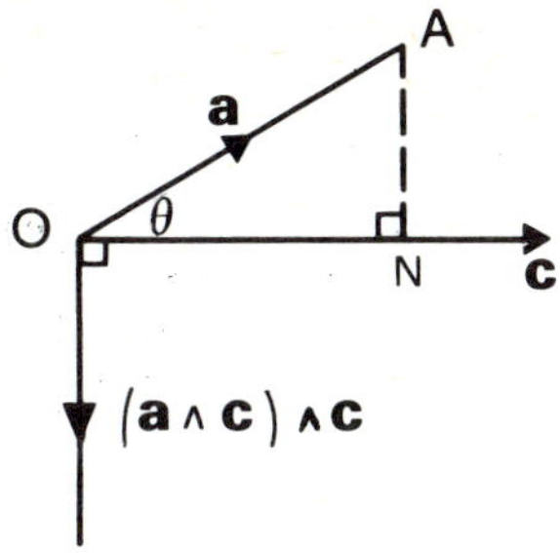

$|(\mathbf{a} \wedge \mathbf{c}) \wedge \mathbf{c}| = \sin\theta$

$$\therefore\ (\mathbf{a} \wedge \mathbf{c}) \wedge \mathbf{c} = \mathbf{AN} = \mathbf{AO} + \mathbf{ON} = -\mathbf{a} + \mathbf{c}\cos\theta$$
$$= (\mathbf{a} \cdot \mathbf{c})\mathbf{c} - \mathbf{a}$$
$$\therefore\ (\mathbf{a} \wedge \mathbf{c}) \wedge \mathbf{c} = (\mathbf{a} \cdot \mathbf{c})\mathbf{c} - (\mathbf{c} \cdot \mathbf{c})\mathbf{a} \qquad (1)$$

as required.

(ii) general case where **a**, **b**, **c** are non-coplanar

Let

$$(\mathbf{a} \wedge \mathbf{b}) \wedge \mathbf{c} = A\mathbf{a} + B\mathbf{b} + C\mathbf{c}. \qquad (2)$$

(*a*) Take the scalar product of (2) with $\mathbf{a} \wedge \mathbf{b}$.

i.e. $(\mathbf{a} \wedge \mathbf{b}) \cdot [(\mathbf{a} \wedge \mathbf{b}) \wedge \mathbf{c}] = C\mathbf{c} \cdot (\mathbf{a} \wedge \mathbf{b})$

but $(\mathbf{a} \wedge \mathbf{b}) \cdot [(\mathbf{a} \wedge \mathbf{b}) \wedge \mathbf{c}] = 0$

$$\therefore\ C = 0.$$

(*b*) Take the scalar product of (2) with $\mathbf{b} \wedge \mathbf{c}$.

i.e. $[(\mathbf{a} \wedge \mathbf{b}) \wedge \mathbf{c}] \cdot (\mathbf{b} \wedge \mathbf{c}) = A\mathbf{a} \cdot (\mathbf{b} \wedge \mathbf{c})$

but
$$[(\mathbf{a} \wedge \mathbf{b}) \wedge \mathbf{c}] \cdot (\mathbf{b} \wedge \mathbf{c}) = [\mathbf{c} \wedge (\mathbf{b} \wedge \mathbf{c})] \cdot (\mathbf{a} \wedge \mathbf{b})$$
$$= -[(\mathbf{b} \wedge \mathbf{c}) \wedge \mathbf{c}] \cdot (\mathbf{a} \wedge \mathbf{b})$$
$$= -[(\mathbf{b} \cdot \mathbf{c})\mathbf{c} - \mathbf{a}] \cdot (\mathbf{a} \wedge \mathbf{b}) \quad \text{from (1)}$$
$$= -(\mathbf{b} \cdot \mathbf{c})[\mathbf{c} \cdot (\mathbf{a} \wedge \mathbf{b})]$$
$$= -(\mathbf{b} \cdot \mathbf{c})[\mathbf{a} \cdot (\mathbf{b} \wedge \mathbf{c})]$$

$$-(\mathbf{b} \cdot \mathbf{c})[\mathbf{a} \cdot (\mathbf{b} \wedge \mathbf{c})] = A\mathbf{a} \cdot (\mathbf{b} \wedge \mathbf{c})$$

i.e.
$$A = -\mathbf{b} \cdot \mathbf{c}.$$

(c) Take the scalar product of (2) with $\mathbf{c} \wedge \mathbf{a}$.

i.e. $$[(\mathbf{a} \wedge \mathbf{b}) \wedge \mathbf{c}] \cdot (\mathbf{c} \wedge \mathbf{a}) = B\mathbf{b} \cdot (\mathbf{c} \wedge \mathbf{a})$$

but
$$\begin{aligned}[(\mathbf{a} \wedge \mathbf{b}) \wedge \mathbf{c}] \cdot (\mathbf{c} \wedge \mathbf{a}) &= [\mathbf{c} \wedge (\mathbf{c} \wedge \mathbf{a})] \cdot (\mathbf{a} \wedge \mathbf{b}) \\ &= [(\mathbf{a} \wedge \mathbf{c}) \wedge \mathbf{c}] \cdot (\mathbf{a} \wedge \mathbf{b}) \\ &= [(\mathbf{a} \cdot \mathbf{c})\mathbf{c} - \mathbf{a}] \cdot (\mathbf{a} \wedge \mathbf{b}) \qquad \text{from (1)} \\ &= (\mathbf{a} \cdot \mathbf{c})[\mathbf{c} \cdot (\mathbf{a} \wedge \mathbf{b})] \\ &= (\mathbf{a} \cdot \mathbf{c})[\mathbf{b} \cdot (\mathbf{c} \wedge \mathbf{a})]\end{aligned}$$

$$(\mathbf{a} \cdot \mathbf{c})[\mathbf{b} \cdot (\mathbf{c} \wedge \mathbf{a})] = B\mathbf{b} \cdot (\mathbf{c} \wedge \mathbf{a})$$

i.e. $$B = \mathbf{a} \cdot \mathbf{c}$$

$$\therefore (\mathbf{a} \wedge \mathbf{b}) \wedge \mathbf{c} = (\mathbf{a} \cdot \mathbf{c})\mathbf{b} - (\mathbf{b} \cdot \mathbf{c})\mathbf{a}.$$

(iii) the case where **a**, **b**, **c** are coplanar is trivial

REFERENCES

7.1. *The Use of Vectors for Finding OH^2, OI^2 in a Triangle*
H. G. Forder 2170, Vol. 34, No. 310, p. 289

7.2. *Triple Vector Product Formula*
G. H. Livens 1381, Vol. 23, No. 255, p. 285
T. G. Room 1384, Vol. 23, No. 255, p. 287
J. L. B. Cooper 2177, Vol. 34, No. 310, p. 296
H. V. Lowry 1109, Vol. 18, No. 229, p. 199
R. J. A. Barnard 1385, Vol. 23, No. 255, p. 288

8 Mechanics

In the following collection of mechanics notes some units have been left as in the originals, for in these cases the change to new units would add no significance.

Although the chapter covers only a limited number of topics, there is a wide range of approaches, varying from basic definitions of physical properties to pure mathematical techniques.

One of the neatest ideas is this addition to the set of graphical treatments of standard constant acceleration formulae.

8.1 Geometrical Proof of $v^2 = u^2 + 2fs$

Pupils sometimes ask whether there is a geometrical proof of this formula, having been shown the v, t diagram proofs of the other formulae for constant acceleration.

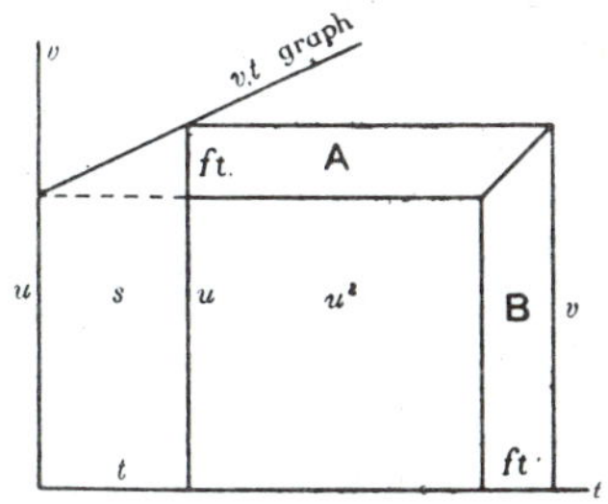

From the diagram, $v^2 - u^2$ = trapezium A + trapezium B. Each of these trapezia has parallel sides u and v, but width ft instead of t as in the s diagram. Each therefore represents fs.

The next note could provide useful practice at graphical interpretation via kinematic ideas.

8.2 A First Lesson on Kinematics—Application of Graphs

The following lesson proved particularly interesting and may give a useful suggestion to any who are anxious to broaden the scope of their work on graphs. It was given to a class who had been drawing graphs of functions and was the first of a series of lessons in which some first ideas on dynamics can be given—preparatory to a fuller course later.

All the pupils had squared paper, and pencils, and were asked to draw the graphs described freehand. Easy scales were suggested and simple numbers were chosen.

The first graph is a time–distance graph. Time is to be represented horizontally, showing 5, 10, 15, 20, 25 minutes on the axis; distance vertically showing 200, 400, . . . , 1000 m on the axis. A pupil sets off from home to go to the dentist; he walks 300 m in the first 5 minutes, runs 600 m in the next 5 minutes, hesitates and stops for 5 minutes, and then thinks better of it and runs home at 150 m/min.

Points raised in a short discussion on this were that the speed is represented by the gradient, and that the 'return' involved only coming back on the distance scale. Several pupils wanted to make their watches go backwards also. But the ideas here were familiar so we did not stop too long on this graph.

Next represent this same journey on a time–velocity graph, taking the same scale for time, and 40, 80, 120 m/min vertically. This proved most interesting. Comparatively few at first saw that the graph consisted of lines parallel to the time-axis. It gave them a real example of a discontinuous graph. They needed coaxing before realising that the return journey could be represented by a negative velocity—one of their first introductions to the idea of 'sense' in measuring a line.

Now for the reverse process, i.e. a time–velocity graph drawn first and the time–distance graph deduced. Taking the same scale for the time–velocity graph, represent the following journey. A pupil starts at 80 m/min for 5 minutes, and then increases his speed uniformly, i.e. 'accelerates' uniformly, for the next 5 minutes to 160 m/min. He stops for 5 minutes and then goes back at a speed which is at first 100 m/min, and decreases uniformly to 80 m/min in the course of the next ten minutes. (Nothing was said about where this landed the pupil.)

After this had been sketched on the board, and the pupils had altered their graphs where wrong—most of them were correct—they were asked to draw the corresponding time–distance graph, using the same scale as before. This proved exciting, and walking round the class one was able to see the progress made. The first 5 minutes was done correctly, but the next 5 minutes was baffling. Not one of their first efforts was correct; that stirred them and three pupils then drew a curve correctly, and after being given some hints half the class got it correct before a solution was given on the board. Then the question was raised as to where the pupil was at the end of the 25 minutes. Working from the time–velocity graph this introduced the idea of average velocity and area under the graph; the distance was calculated, and the pupils had a check on the accuracy of their freehand time–distance graph.

They had in these graphs met the idea of the gradient of the tangent to a curve representing velocity, as well as of the area under a graph representing distance. There was just time to draw a curved graph, time–velocity, given as speedometer readings of a car at various times, to show the importance of this latter idea for 'irregular' motion.

The above introduction to kinematics concentrates on rapidly developing a technical skill, but much of mechanics is concerned with building mathematical models, and the initial assumptions can have fundamental effects. The two following notes consider different starting assumptions for the same problem. Both simplify the real situation—a necessary feature of any model, and both give useful advice on how the simplifications might be chosen.

8.3 A Walk in the Rain

Rain is falling in a north wind at an angle α to the vertical with a velocity y, and a man is to make a journey through the rain in a direction at an angle β with north. At what speed must he walk to keep as dry as possible, assuming the conditions will remain constant?

The amount of rain he encounters, R, will be equal to (time taken) $\times$ (area exposed normally to direction of rain relative to himself) $\times$ (flow of rain relative to himself). We shall take the man to be a sphere of unit radius, and the flow of rain and the distance to be travelled as units. Let x be the man's speed, and z the velocity of the rain relative to the man. From the diagram (see opposite) we get

$$z = (x^2 + 2xy \sin\alpha \cos\beta + y^2)^{\frac{1}{2}},$$

and since, at constant density, flow is proportional to velocity, it follows that

$$R = \frac{1}{x} \, . \, \pi \, . \, \frac{z}{y}$$

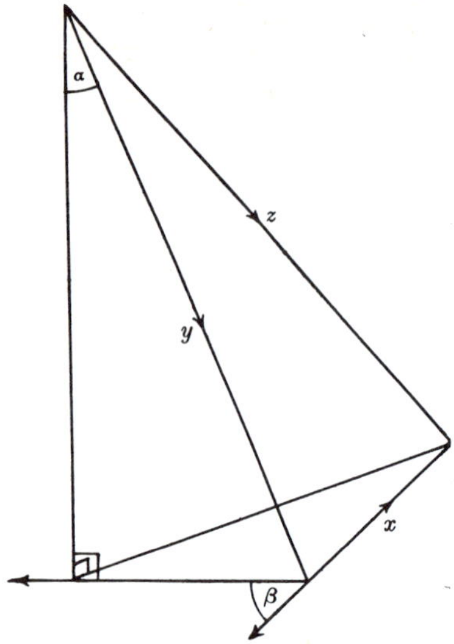

$$= \frac{\pi(x^2+2xy\sin\alpha\cos\beta+y^2)^{\frac{1}{2}}}{xy}.$$

Hence $$\frac{dR}{dx} = \frac{-\pi(x\sin\alpha\cos\beta+y)}{x^2(x^2+2xy\sin\alpha\cos\beta+y^2)^{\frac{1}{2}}}$$

$$= 0 \quad \text{when} \quad x = \frac{-y}{\sin\alpha\cos\beta}.$$

Thus when $\beta \leqq 90°$, the best the man can do is go as fast as possible, but when $\beta > 90°$ he must make the southerly component of his speed equal to y . cosec α.

The solution for forms more closely resembling the human form is left to the reader: in these the areas presented by the man will vary with his speed, and a more complex expression for R result. The rule given above, however, is probably a good guide to practice.

Alternatively, we could try as starting assumptions:

(i) The wetting power of the rain is proportional to the *square* of the relative velocity.

(ii) The man's shape is that of a right block, height a, breadth b, depth (back to front) c.

The only case considered is when the rain falls vertically with velocity u and the man moves forward with velocity v.

Let $v = u\tan\theta$.

Square of relative velocity $= u^2\sec^2\theta$.

Exposed surface $= (a\sin\theta + c\cos\theta)b$.

Time for distance s $= \dfrac{s}{v} = \dfrac{s}{u}\cot\theta$.

The amount of wetting is proportional to

$$u^2\sec^2\theta(a\sin\theta + c\cos\theta)\frac{s}{u}\cot\theta$$

$$= us\left(\frac{a}{\cos\theta} + \frac{c}{\sin\theta}\right).$$

This will be a minimum if

$$\frac{a\sin\theta}{\cos^2\theta} - \frac{c\cos\theta}{\sin^2\theta} = 0$$

i.e. if $$\tan^3\theta = \frac{c}{a}$$

or $$v^3 = \frac{c}{a}u^3.$$

It is interesting to see that the *short fat man*, for whom c/a is greater than for the tall thin man, *must move faster* than the other.

Here the assumptions were explicit and their effects easily traceable. The next two results explore the effects of the implicit assumptions of the models.

8.4 Implicit Assumptions in Standard Problems

Question. Two particles m_1, m_2 connected by a light inelastic string rest on a smooth horizontal plane. The string is slack and m_2 is projected with velocity u away from m_1. What are the velocities when both particles are in motion?

Solution. Let v be the common velocity after m_1 is jerked into motion; by the conservation of momentum, $(m_1 + m_2)v = m_2u$.

Query. Is it possible to justify the assumption above that m_1 and m_2 will have a common velocity? Most pupils, when asked to think about this, say that m_1 will have a greater velocity than m_2.

Answer. In practice the string must be slightly elastic; m_1 cannot instantaneously acquire a velocity v. Let l be the natural length of the string and λ its (very large) modulus of elasticity.

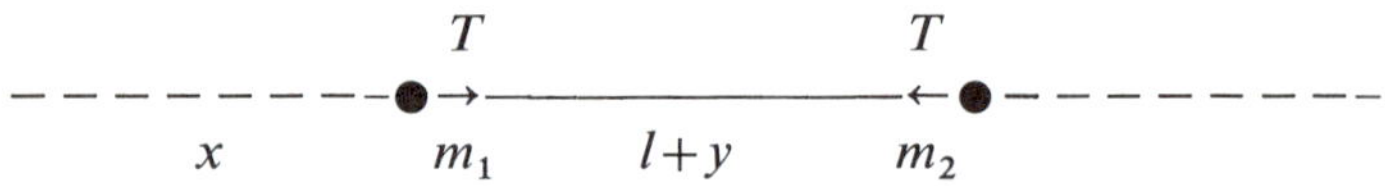

Then $$T = m_1\ddot{x} = -m_2(\ddot{x}+\ddot{y}) = \lambda y/l.$$

Eliminating $\ddot{x}$, we have $\ddot{y} = -\mu y$, where $\mu = \lambda(m_1+m_2)/lm_1m_2$.

The initial conditions are: $t = 0,\ x = 0,\ y = 0,\ \dot{x} = 0,\ \dot{y} = u$.

Thus $\dot{y} = u\cos t\sqrt{\mu}$ and $y = (u/\sqrt{\mu})\sin t\sqrt{\mu}$,

whence $\ddot{x} = (\lambda u/m_1 l\sqrt{\mu})\sin t\sqrt{\mu}$,

and $\dot{x} = \lambda u(1-\cos t\sqrt{\mu})/m_1 l\mu = v(1-\cos t\sqrt{\mu})$.

The velocity of m_1 is $v - v\cos t\sqrt{\mu}$,

the velocity of m_2 is $v+(u-v)\cos t\sqrt{\mu}$,

and the period $2\pi/\sqrt{\mu}$ is very small.

Time	*Velocity of m_1*	*Velocity of m_2*	*String*
0	0	u	natural length;
$\pi/2\sqrt{\mu}$	v	v	greatest extension;
$\pi/\sqrt{\mu}$	$2v$	$2v-u$	natural length.

When the extension of the string is greatest, the masses have the common velocity v, and this state of affairs will continue if the string is permanently deformed. In this case the assumption is justified.

If the string is elastic, when the velocities are $2v$, $2v-u$ and the string is at its natural length, it will go slack and the simple harmonic motion ceases, so that those velocities are maintained, thus confirming the pupils' guesses. Note that $2v-u = (m_2-m_1)u/(m_1+m_2)$, and so m_2 will be jerked backwards if $m_1 > m_2$.

In practice the actual behaviour will probably be somewhere between the above two cases.

Or, inspired by a school examination paper,

'A cyclist travels on a certain journey at 8 miles per hour against a head wind. On his return journey the wind is still blowing at the same rate, and his speed is 20 miles per hour. Calculate the speed of the wind.'

At first sight, and without reflection, the young candidates may well have been tempted to give the answer as 6 m.p.h., and to regard the cyclist as always travelling at 14 m.p.h. relative to the moving air; but consideration of what would happen if the head wind freshened to 15 m.p.h. should have warned them against this obvious trap.

In the absence of explicit guidance the following assumptions seem reasonable: that frictional resistances can be neglected, that the air resistance is proportional to the square of the relative velocity, and that the cyclist works at a uniform rate. Then, if the speed of the wind is x m.p.h., the resistances in the two directions are in the ratio $(8+x)^2 : (20-x)^2$, and we are led to the equation

$$8(8+x)^2 = 20(20-x)^2,$$

whence $$x = \tfrac{1}{3}(116-28\sqrt{10}) = 9{\cdot}15\ldots$$

It would be interesting to know if this result was generally obtained.

In general, if the cyclist's *actual* speeds against and with the wind are u and v, it will be found that (with these assumptions) his *relative* speeds in the two directions are $\sqrt{v}\,.\,(u+v)/(\sqrt{u}+\sqrt{v})$ and $\sqrt{u}\,.\,(u+v)/(\sqrt{u}+\sqrt{v})$. In setting such a question, calculations are simplified if v/u is taken to be an exact square. Thus, if a cyclist is taken to travel at 4 m.p.h. against the wind and 36 m.p.h. with the wind, the relative velocities are found to be 30 m.p.h. and 10 m.p.h., and the speed of the wind to be exactly 26 m.p.h.

Sometimes, an apparently incorrect model can give 'the right answers'. The next article is particularly thought provoking in this respect.

8.5 Some Remarks on the Coefficient of Restitution

The coefficient of restitution e which occurs in the discussion of the impact of elastic bodies can be defined in two ways:

(i) by Newton's experimental law of impact;

(ii) by the theory (due to Poisson) which defines e as the ratio of the impulse of restitution to the impulse of compression.

It is the purpose of this note to examine each of these definitions and to set out some of the assumptions implied in them. As a framework on which to hang the discussion, I take two examples.

Example (*a*). A body of mass m_1 moving with velocity u_1 impinges directly on a body of mass m_2 moving with velocity u_2. Find the resulting velocities.

Example (*b*). A sphere of mass m_1 is suspended from a fixed point by an inextensible string; a second small sphere of mass m_2 moving with velocity V downwards in a direction making an angle α with the vertical impinges directly on the first sphere. Find the velocities of the spheres after impact.

For the sake of completeness we now quote Newton's experimental law: 'When two bodies A and B impinge, the velocity after impact of A, relative to B, is equal to $-e$ times the velocity before impact of A relative to B; the velocities being measured along the common normal.' Here it is tacitly assumed that e is independent of the velocities before impact and the shapes and masses of the bodies. When e is defined by (ii) the same assumptions are made; there is, however, one important difference.

In definition (i) we assume that the only impulses acting on the bodies are the mutual impulses exerted between them. This is not done when defining e from (ii). Problems in impact can be divided into two classes, those in which the mutual impulses are the only ones acting on the bodies at the moment of impact, and those in which other impulses may act in addition to the mutual impulses. Example (*a*) is typical of the first class, example (*b*) of the second; the impulsive tension in the string is of course the additional impulse.

Let us consider further the contention made in the last paragraph. If in defining e from (i) we do not assume that the mutual actions between the bodies are the only impulses acting, then e will depend on the other impulses, and for the impact of two bodies of given materials e will depend on the conditions of the particular problem discussed. Thus suppose that in examples (*a*) and (*b*) the spheres of mass m_1 are made of steel and those of mass m_2 are made of glass, then there appears to be no valid reason for supposing that the coefficients of restitution in the two examples are equal (as defined by Newton's law). We cannot say that the coefficient of restitution between steel and glass is, say, $\frac{3}{4}$, but only that for the particular problem discussed the coefficient of restitution between steel and glass is $\frac{3}{4}$. Therefore since e is a constant only for mass, velocity, shape, etc., but is to vary with the problem for given materials, there is little use in its introduction. The value of any physical experimental constant lies in the fact that once found it can be used to forecast the results of given sets of conditions. If e is to be of value, then it must depend solely on the materials of the impinging bodies and not on the class of the problem.

In defining e according to (i) it is assumed that the only impulses acting on the bodies are the mutual impulses between them. The conclusion is that if we adopt method (i) we are justified in using it to solve problems of type (*a*) but not of type (*b*). It should here be mentioned that if we confine our attention to problems of type (*a*), then taking (i) as our definition we can establish (ii) and *vice versa*. This is proved in any elementary textbook. In this class of problem neither definition has therefore any merit over the other.

Examples of type (*b*) can only be solved logically by defining e as in (ii). It is this point which has prompted this note. Most intermediate textbooks, and certainly most schoolbooks, offend in this respect. The curious point, however, is that the erroneous application of Newton's law gives the correct result in most of the cases which the writer has examined. The solution of example (*b*) is given below.

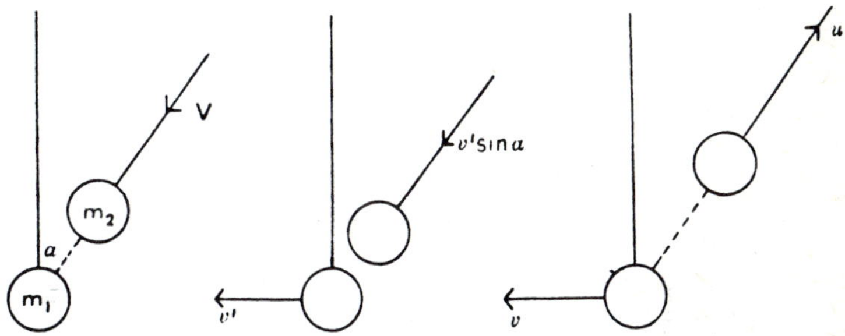

Before impact End of period of compression After impact

At the end of the period of compression let v' be the velocity of m_1; v' is of course horizontal. Since the relative velocity of the spheres is now zero, the velocity of m_2 is $v' \sin\alpha$ as shown.

Let v and u be the velocities of m_1 and m_2 after impact.

Let I be the impulse of compression; then eI is the impulse of restitution. Considering m_2 we have

$$I = m_2(V - v' \sin\alpha), \qquad (1)$$

$$eI = m_2(u + v' \sin\alpha). \qquad (2)$$

Considering m_1 and resolving horizontally to avoid the impulsive tension in the string, we have

$$I \sin\alpha = m_1 v', \tag{3}$$

$$eI \sin\alpha = m_1(v - v'). \tag{4}$$

We have four equations to determine the four unknowns I, v', u and v, which give

$$u = V(em_1 - m_2 \sin^2\alpha\}/(m_1 + m_2 \sin^2\alpha),$$

$$v = V\{(1+e)m_2 \sin\alpha\}/(m_1 + m_2 \sin^2\alpha).$$

In a textbook the solution is given thus:

The momentum equation in the horizontal direction is

$$m_2 V \sin\alpha = m_1 v - m_2 u \sin\alpha; \tag{5}$$

applying Newton's law along the line of centres,

$$u + v \sin\alpha = eV. \tag{6}$$

Solving these equations we have u and v as above.

The system (1)–(4) does imply (5) and (6), so perhaps Newton's law does have a range of applicability wider than that suggested at the beginning of this article.

These foregoing articles deal with basic principles, but much school mechanics deals with technique, after the mathematical model has been formulated. The next article contrasts particularly with the previous discussion.

8.6 An Elementary Collision Problem

The following problem provides a good example of the algebra that arises in collision problems and may be of interest to anyone with a few hours to spare. It may at first sight appear to be very simple, but the use of the word 'hours' is correct.

Four exactly similar balls A_1, A_2, A_3, A_4 are at rest at equal distances apart in a smooth infinite horizontal groove. A_1 is then projected towards A_2 with velocity v. If e is the coefficient of restitution for any pair of balls, what is the subsequent behaviour of the system?

It will be found that the value of e is important, and the different cases that arise are as follows:

I. $e \leqq 3-2\sqrt{2}$.

The balls do not finally separate, i.e. no time t can be found such that any ball does not take part in a collision after time t. The velocity of each ball tends to become $v/4$.

II. $e > 3-2\sqrt{2}$.

After a finite number of collisions, the balls all separate and do not meet again. If $N(1,2)$ is the number of collisions between A_1 and A_2, etc., these numbers may be found as follows. Let θ be defined by $\cos\theta = (1-e)^2/4e$, $0 < \theta < \frac{1}{2}\pi$, and let n be the positive integer given by $\dfrac{\pi}{n} \leqq \theta < \dfrac{\pi}{n-1}$. Then if n is even, the numbers of collisions are

$$N(1,2) = N(2,3) = N(3,4) = n-1,$$

and if n is odd, the numbers are

$$N(1,2) = n, \quad N(2,3) = n-1, \quad N(3,4) = n-2.$$

The final state is particularly interesting when $\theta = \pi/n$. If n is even, A_1, A_2, A_3 finally have the same velocity $v(1-e^n)/4$, less than that of A_4, $v(1+3e^n)/4$. If n is odd, A_2, A_3, A_4 finally have the same velocity $v(1+e^n)/4$, greater than that of A_1, $v(1-3e^n)/4$.

It is easily seen that n decreases rapidly from infinity as e increases from the value $3-2\sqrt{2}$ or $\cdot 1716$. Thus, when $e = \cdot 172$, $n = 38$, when $e = \cdot 2$, $n = 5$, and when $e \geqq 2-\sqrt{3} = \cdot 268$, $n = 3$.

The above results are obtained under the assumption that consecutive collisions between A_2 and A_3 are separated by a collision between A_1 and A_2 and another between A_3 and A_4. This is most probably true, but one needs to prove it generally.

8.7 Theorems on Centroids

Here are two results which do not normally appear with the standard theorems for simple plane laminae.

I Irregular quadrilateral replaced by three equal particles (Fig. 1).

If J, K are points on the diagonals (which intersect at I) such that $JC = AI$ and $KB = DI$, the centre of gravity of the quadrilateral coincides with that of three equal particles situate at I, J, K.

FIG. 1

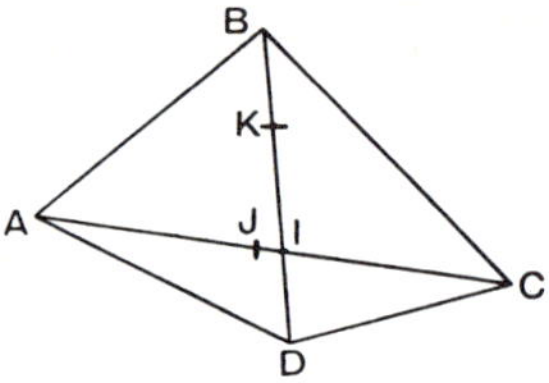

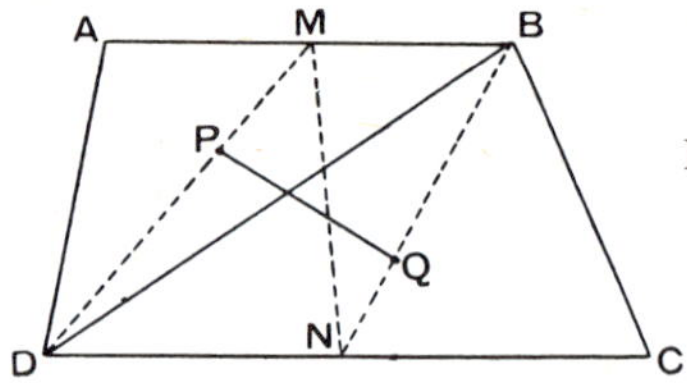

FIG. 2

II Trapezium replaced by two particles of masses proportional to partial areas. (Fig. 2).

Let a, b be the lengths of the parallel sides, h the distance between them. The area can be replaced by particles P, Q of masses proportional to a, b at the centres of gravity of the triangles ABD, CBD.

Thus taking moments about CD,

$$(a+b)\bar{y} = a \,.\, \tfrac{2}{3}h + b \,.\, \tfrac{1}{3}h, \text{ etc.}$$

This is slightly simpler than taking particles at the angular points, and also more obvious, as P, Q are actual centres of gravity of the two parts. Also, it lends itself to graphical construction, since the centre of gravity is at the intersection of MN and PQ.

8.8 Further Centre of Gravity Problems

I To find the centre of gravity of an orange section of a sphere

Let the semi-circular faces contain an angle 2θ. If an elemental 'orange section' has a centre of gravity at a distance x from its straight edge, the whole body is statically equivalent to a uniform arc of radius x subtending 2θ at its centre; and the c.g. of this is $\dfrac{x\sin\theta}{\theta}$ from the centre of the sphere.

In the case of a hemisphere, $\dfrac{x\,.\,1}{\pi/2} = \dfrac{3}{8}a$. Hence the required distance is $\dfrac{3\pi}{16}\dfrac{a\sin\theta}{\theta}$.

As a corollary, letting $\theta \to 0$, the c.g. of a semi-circular lamina with density proportional to the distance from the bounding diameter is $\dfrac{3\pi}{16}a$ from that diameter.

And now two interesting alternative approaches to the same problem:

II To find the c.g. of a pyramid without integration

A square pyramid can be cut into slices in two distinct ways—by planes parallel to the base, or by planes parallel to the plane through the apex and a pair of opposite corners of the base. The first method of slicing produces thin square plates whose areas are proportional to the squares of their linear dimensions, and therefore to the squares of their distances from the apex. The second produces triangular plates whose areas are proportional to the squares of their heights. Pair off each of these triangles with its congruent partner on the opposite side of the first central cut. The two together have a c.g. on the central axis of the pyramid at one-third of their height above the base. Now if A is the apex of the pyramid and N the foot of the perpendicular from A to the square base, we have two alternative sets of masses with their centres of gravity on AN; one is a series of masses with magnitudes proportional to the squares of their distances from A, running from zero at A to a maximum at N; the other is a similar series of masses, but now the magnitudes are proportional to the squares of their distances from N, and the zero is at N and the maximum at G, where $AG = 2 \,.\, AN/3$. We may assume that their common c.g., the c.g. of the whole pyramid, is in the same relative position with respect to these two series of masses. Hence, if it is at K, we have $AK/KN = NK/KG = x$, say, from which we obtain at once $3+2x = x^2$, and $x = 3$.

III To find the centroid of a pyramid without integration

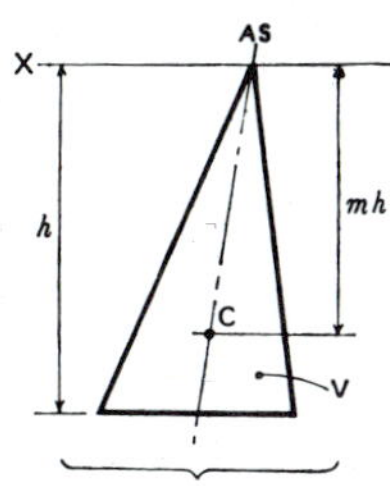

FIG. 1

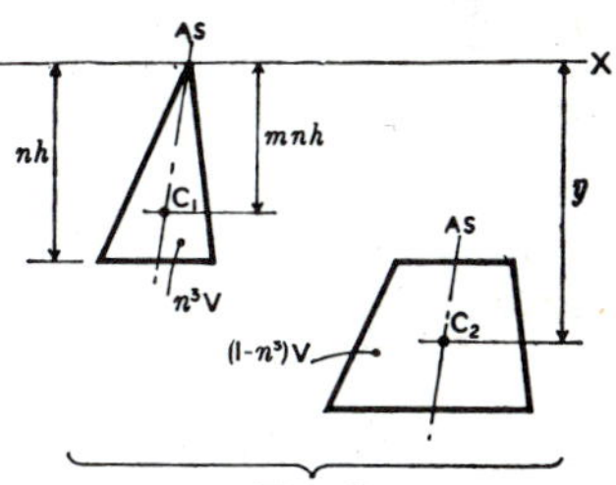

FIG. 2

Fig. 1 represents a pyramid of height h, volume V and centroid C.

Fig. 2 shows the pyramid cut into two parts by a plane parallel to the base and distant nh from the apex. The volumes and centroids of these parts are indicated. XX is a line parallel to the base.

We require to find the value of m.

Taking moments about the line XX, we have

$$V \, . \, mh = n^3 V \, . \, mnh + \bar{y} \, . \, (1-n^3)V.$$

$$\therefore \qquad mh(1-n^4) = \bar{y}(1-n^3).$$

$$\therefore \qquad m = \frac{\bar{y}(1-n^3)}{h(1-n^4)}$$

$$= \frac{\bar{y}(1+n+n^2)}{h(1+n)(1+n^2)} = f(n), \quad \text{say.}$$

Now as $n \to 1$, $\bar{y} \to h$, so that m, being in fact independent of n, is equal to $\lim_{n \to 1} f(n) = \frac{3}{4}$, i.e. the centroid of the pyramid is distant $\frac{3}{4}h$ from the apex.

8.9 Moments of Inertia

In this section there are just two examples of Moment of Inertia calculations. The first is included simply because it is a very neat, concise solution to a problem which one suspects, at first sight, to be complicated. The second has, beside the virtue of neatness, echoes of the shearing technique found in several new mathematics courses.

I Moment of inertia of a triangular lamina

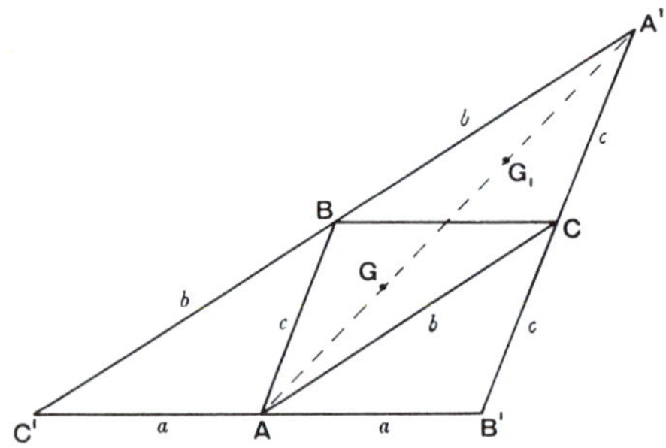

Let the moment of inertia of a uniform triangle about its centre of gravity be Mk^2, M being its mass and k^2 a function of its sides a, b and c.

Then the moment of inertia of the triangle $A'B'C'$ is $16Mk^2$, and this is the sum of the moments of inertia about G of its four parts, so that

$$16Mk^2 = Mk^2 + \Sigma(Mk^2 + M \,.\, GG_1{}^2).$$

Hence

$$\begin{aligned} 12k^2 &= \Sigma GG_1{}^2 \\ &= \tfrac{1}{9}\Sigma A'A^2 \\ &= \tfrac{1}{9}\Sigma(2b^2+2c^2-a^2) \\ &= \tfrac{1}{3}\Sigma a^2. \end{aligned}$$

Thus the moment of inertia about its centre of gravity of a uniform triangular lamina of mass M with sides a, b and c is

$$M(a^2+b^2+c^2)/36.$$

II Inertia of a conical shell

The moment of inertia of a right hollow cone (or conical shell) of base radius, a, and mass m, about the axis of symmetry, is $\tfrac{1}{2}ma^2$.

This being the same result which one obtains for the solid cylinder of radius a, mass m, about the same axis, suggests that the cylinder may be obtained from the cone by translations parallel to the axis. The following self-evident illustrations show this to be the case.

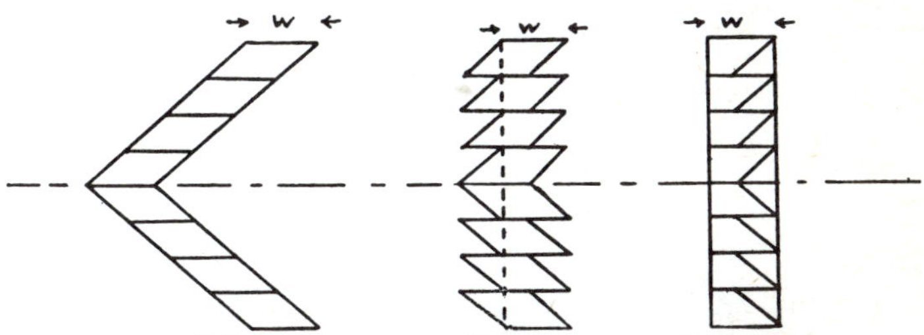

Pure mathematical technique is a prominent feature of traditional applied mathematics, and the following four notes are representative of techniques with wide applicability.

8.10 Proof of the Formula for Acceleration of a Point Moving with Uniform Speed in the Circumference of a Circle

Consider the motion of a point moving with uniform speed v in the perimeter of a regular polygon of n sides whose circumcircle has radius r.

Let O be the centre of the circle, and A, B, C three successive vertices of the polygon.

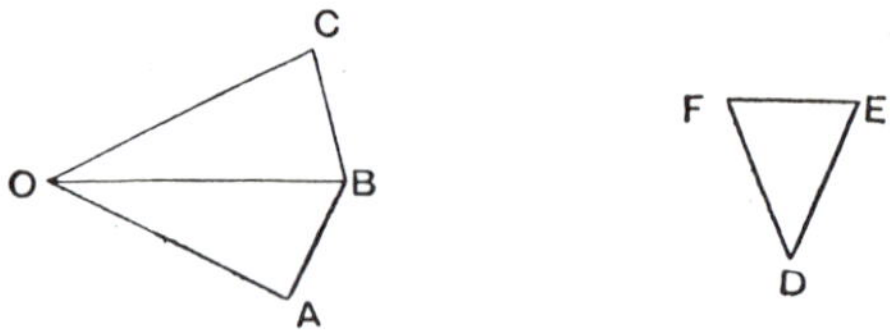

Let DE, DF represent in magnitude and direction the velocities in AB, BC respectively. Then EF represents on the same scale, in magnitude and direction, the change of velocity that takes place at B.

Hence that change is $= v \cdot \dfrac{EF}{DE} = v\dfrac{AB}{OA}$ since the triangles DEF, OAB are similar.

If t be the time taken to move from A to B, the average rate of change of velocity is $\dfrac{vAB}{tOA} = \dfrac{v^2AB}{vt \cdot OA} = \dfrac{v^2}{OA}$, since $vt = AB$.

Now let $n \to \infty$. In the limit the polygon becomes a circle of radius $OA = r$.

Hence the acceleration of the circle is $\dfrac{v^2}{r}$.

Also EF, being equally inclined to DE and DF is parallel to BO. i.e. the acceleration is towards the centre.

8.11 The Elementary Theorems in Analytical Statics

The sum of the moments of forces $P_1, P_2, \ldots$ acting along the lines $p_1 - x\cos\alpha_1 - y\sin\alpha_1 = 0$, etc., about (h, k) is

$$\Sigma P_1(p_1 - h\cos\alpha_1 - k\sin\alpha_1),$$

or

$$\Sigma P_1 p_1 - h\Sigma P_1\cos\alpha_1 - k\Sigma P\sin\alpha_1,$$

which is usually written

$$G + kX - hY. \qquad (1)$$

Thus the moment of the system about any point is known provided that X, Y and G are known. We say that the system has coordinates X, Y, G, and refer to it as the system (X, Y, G).

A couple is a system of coplanar forces having (a) constant moment, (b) zero component in all directions. The necessary and sufficient conditions for (X, Y, G) to be a couple are easily proved to be $X = Y = 0$, $G \neq 0$, so that $(0, 0, G)$ is a couple.

The expression (1) can be written

$$\sqrt{(X^2+Y^2)}\left\{\frac{G+kX-hY}{\sqrt{(X^2+Y^2)}}\right\},$$

unless $X = Y = 0$, and is the moment of a single force $\sqrt{(X^2+Y^2)}$, acting along $G+yX-xY = 0$.

So *a system of coplanar forces can be reduced to a single force or a couple.*

By consideration of (i) it appears that the system (X_1, Y_1, G_1) plus the system (X_2, Y_2, G_2) has the same moment about all points as the system $(X_1+X_2, Y_1+Y_2, G_1+G_2)$.

Hence the addition theorem

$$(X_1, Y_1, G_1)+(X_2, Y_2, G_2) = (X_1+X_2, Y_1+Y_2, G_1+G_2).$$

Consequently, since $(0, 0, G)+(0, 0, -G) = (0, 0, 0)$, *two couples of equal and opposite moment are in equilibrium*, and since

$$\Sigma(0, 0, G) = (0, 0, \Sigma G),$$

the resultant of a number of coplanar couples is a couple of moment equal to the sum of the moments of the couples.

Moreover, since

$$(X, Y, G_1)+(0, 0, G_2) = (X, Y, G_1+G_2),$$

the resultant of a force and a couple is a force parallel to the first force (the direction in both cases making an angle $\arctan Y/X$ with the axis of x).

Again,

$$(X, Y, G) = (X, Y, 0)+(0, 0, G).$$

Therefore *a system of coplanar forces may be reduced to a force acting at a given point together with a couple* (the origin being taken at the given point).

8.12 The Reduction of the Mass-Accelerations and the Momentum of a Rigid Lamina

For certain points, A, of a rigid lamina, and under certain dynamical circumstances, $L_A = I_A\dot{\omega}$, where L is the sum of the moments of the mass-accelerations of all particles about A. What *are* these points and circumstances?

Let A be a point (x_1, y_1) of a rigid lamina, relative to axes fixed in space. Let any point of the lamina have coordinates (x, y), or (x_2, y_2) relative to A, so that

$$x = x_1 + x_2 \quad \text{and} \quad y = y_1 + y_2.$$

Now the mass-accelerations of the system of particles of the body form a system of localised vectors and hence, like forces, can be reduced to (*a*) a single mass-acceleration acting at A, and (*b*) a mass-acceleration couple.

(*a*) The single mass-acceleration

$$= \Sigma(m\ddot{x}, m\ddot{y}),$$

$m\ddot{x}$ and $m\ddot{y}$ being the two components of the mass-acceleration of any particle m of the lamina,

$$= \frac{d^2}{dt^2}\Sigma(mx, my)$$

$$= \frac{d^2}{dt^2}(M\bar{x}, M\bar{y}),$$

where $(\bar{x}, \bar{y})$ is the centre of gravity of the body,

$$= (M\ddot{\bar{x}}, M\ddot{\bar{y}}),$$

which is the mass-acceleration of a particle of mass M situated at the centre of gravity of the system, G, and moving with the acceleration of G.

(*b*) The mass-acceleration couple is the sum of the moments of the mass-accelerations of the particles about A

$$\begin{aligned}
&= \Sigma m(x_2\ddot{y} - y_2\ddot{x}) \\
&= \Sigma m\{x_2(\ddot{y}_1 + \ddot{y}_2) - y_2(\ddot{x}_1 + \ddot{x}_2)\} \\
&= \ddot{y}_1\Sigma(mx_2) - \ddot{x}_1\Sigma(my_2) + \Sigma m(x_2\ddot{y}_2 - y_2\ddot{x}_2) \\
&= M(\bar{x}_2\ddot{y}_1 - \bar{y}_2\ddot{x}_1) + I_A\dot{\omega},
\end{aligned}$$

where $(\bar{x}_2, \bar{y}_2)$ is the centre of gravity relative to A.

This reduces to $I_A\dot{\omega}$ in three cases,

(i) A is the centre of gravity, since then $\bar{x}_2 = \bar{y}_2 = 0$;

(ii) A is moving with a constant velocity or is a fixed point, since $\ddot{x}_1 = \ddot{y}_1 = 0$, which includes the special case of rotation about a fixed axis; or

(iii) the acceleration of A is parallel to AG, since then $\dfrac{\ddot{y}_1}{\ddot{x}_1} = \dfrac{\bar{y}_2}{\bar{x}_2}$.

This sometimes includes the case when A is at the instantaneous centre.

The equations of motion can now be obtained by expressing the equivalence of the forces and the reduced mass-accelerations; in particular, $L_A = I_A\dot{\omega}$ follows by taking moments about A.

Momentum can be treated in a similar way. For the momenta of the particles again form a system of localised vectors which can be reduced to (*a*) a single momentum at A, the linear momentum of the lamina, and (*b*) a momentum-couple called the angular momentum of the lamina about A.

(*a*) The linear momentum

$$= \Sigma(m\dot{x}, m\dot{y})$$
$$= (M\dot{\bar{x}}, M\dot{\bar{y}}).$$

(*b*) The angular momentum about A

$$= \Sigma m(x_2\dot{y} - y_2\dot{x})$$
$$= \Sigma m\{x_2(\dot{y}_1 + \dot{y}_2) - y_2(\dot{x}_1 + \dot{x}_2)\}$$
$$= M\bar{x}_2\dot{y}_1 - M\bar{y}_2\dot{x}_1 + I_A\omega.$$

This reduces to $I_A\omega$ in the three cases:

(i) A is the centre of gravity, since then $\bar{x}_2 = \bar{y}_2 = 0$.

(ii) The velocity of A is zero, since than $\dot{x}_1 = \dot{y}_1 = 0$. Hence A is a fixed point or is at the instantaneous centre.

(iii) The velocity of A is parallel to AG, since then $\dfrac{\dot{y}_1}{\dot{x}_1} = \dfrac{\bar{y}_2}{\bar{x}_2}$.

Momentum equations can now be obtained by expressing the equivalence of the impulses and the change of momentum of the lamina.†

† This treatment is developed fully in
Palmer, A. H. G. and Snell, K. S., *Mechanics*, University of London Press, 1935.

And on the same topic,

8.13 A Geometric Sideline on 8.12

Let a rigid lamina, of mass M and centre of mass G, move in its own plane. At a given instant let ω be the angular velocity and I the instantaneous centre; then the velocity of G is $GI\,.\,\omega$ in the direction GX, say. With GX, GI as axes of coordinates, let P be the point (x, y) in the lamina. The moment of inertia about G being Mk^2, that about P is $M(x^2+y^2+k^2)$, and the angular momentum about P is $M(GI\,.\,y+k^2)\omega$, the sense of measurement being the same as that of ω.

Hence, if the angular momentum about P is equal to the product of ω and the moment of inertia about P,

$$x^2+y^2-GI\,.\,y = 0.$$

Thus the point P lies on the circle which has GI as diameter.

This result, even if not new, does not seem to be well known. It may be useful to teachers in overcoming the tendency of beginners (and others!) to take the angular momentum about any point to be the product of the angular velocity and the moment of inertia about that point.

8.14 A Problem on Friction

Standard problems on friction sometimes contain hidden subtleties. Here is one from an examination question.

A bead A, of weight W; is threaded on a rough circular wire, whose centre is O and whose plane is vertical; the bead is attached to a light inextensible string which passes through a fixed smooth hook B at the upper end of the vertical diameter and carries a weight W'; the angle BOA is denoted by 2θ. When the bead is on the point of slipping downward, $\theta = \alpha$; when it is on the point of slipping upward $\theta = \beta$. Prove that if $\alpha+\beta = \dfrac{\pi}{2}$, the coefficient of friction between the bead and the wire must be $\tan(\alpha-\frac{1}{4}\pi)$. Prove also that

$$W' = W(\sin\alpha+\cos\alpha).$$

The required results are easily obtained, either from the composite Fig. 1, in which OH, OK are parallel to BA_1, BA_2 and triangles OLH and OLK are taken as triangles of forces, or from Fig. 2 by resolving along the tangent and radius at A_1 and A_2, *provided we assume that the*

FIG. 1

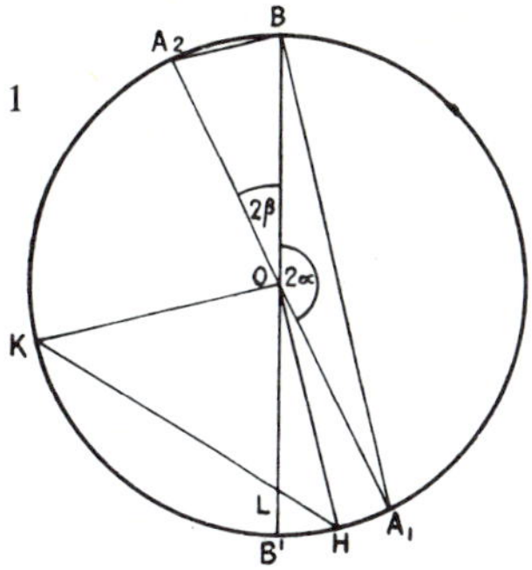

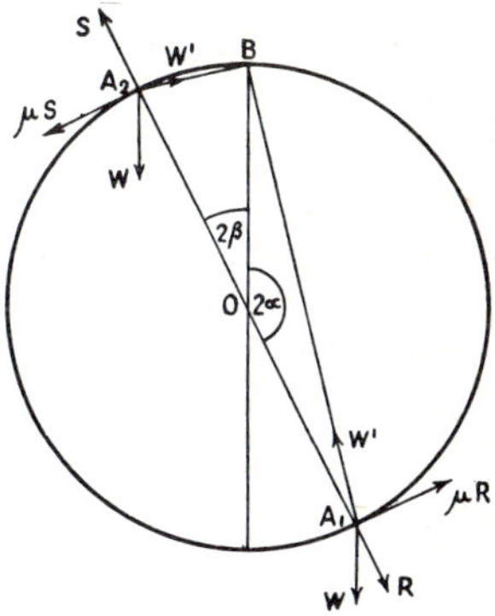

FIG. 2

normal reaction R acts outwards. The question arises—can R act inwards? Can we, in Fig. 1, find a point L' on BB' such that HL' and KL' are equally inclined to A_1A_2 but on opposite sides?

Now the locus of a point P such that HP and KP are equally inclined to A_1A_2 but on opposite sides is very easily shown by coordinate geometry to be a rectangular hyperbola through H and K, whose centre bisects HK and whose asymptotes are parallel and perpendicular to A_1A_2. It would appear likely that in favourable circumstances this hyperbola might cut BB' in two points, each a possible position of L'. (See Fig. 3.)

If the equations of resolution are revised for the change in sense of R, we obtain, after elimination,

$$X^2 \cos 2\beta - X(\sin 3\beta + \cos 3\beta) + \sin 4\beta = 0$$

$$\text{where} \quad X \equiv \frac{W'}{W}$$

and this equation gives real values for X if $4s^3 - 5s + 1 \geq 0$ where $s \equiv \sin 2\beta$. Now $4s^3 - 5s + 1 = 0$ when $s = 1$, $0{\cdot}207$ or $-1{\cdot}2$ approx., and hence we obtain real values of X if $0 < \sin 2\beta \lesssim 0{\cdot}2$. Taking as an example $\beta = 4°$, or $\sin 2\beta \doteqdot 0{\cdot}14$, we get $X \doteqdot 0{\cdot}88$ or $0{\cdot}32$; i.e., $W' \doteqdot \frac{8}{9}W$ or $\frac{1}{3}W$.

This means that, in Fig. 3, if β were 4°, the hyperbola would cut BOB' in two points, say L_1' and L_2' which make $\theta = \phi$ and such that $HO \doteqdot \frac{8}{9}OL_1'$ and $HO \doteqdot \frac{1}{3}OL_2'$.

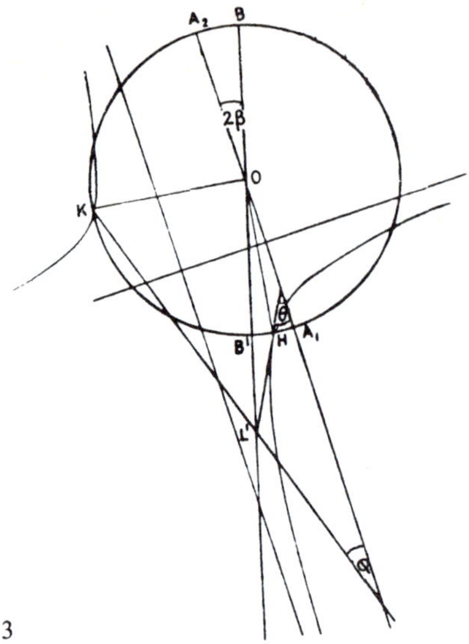

FIG. 3

Thus if β is less than about 6° there are, in addition to the expected solution, two solutions in which the reaction on the bead in the lower position acts inwards, and in which the required results do not hold.

There is no subtlety in the next note! Despite the historic units it provides a very simple connection between μ and braking efficiency.

8.15 Braking Efficiency in Motor Cars

The Brake-Testing Service for motor cars gives the vehicle a certificate of 100% efficiency if it can be stopped in 30 feet from a speed of 30 miles per hour. This will occur when the coefficient of friction is almost unity. For, from $Fs = \mu Ws = Wv^2/2g$,

$$\mu = v^2/2gs = 44^2/64{\cdot}4 \;.\; 30$$

$$= \tfrac{1936}{1932} = 1{\cdot}00.$$

The braking experts consider that an efficiency of 70% is 'excellent', while even 50% is 'good'. It appears, then, that $\mu = \frac{1}{2}$ is average for a modern car, and that 'braking efficiency' is numerically equal to the coefficient of friction.

The principles of conservation and energy and momentum are probably the two most used and most mis-used principles of physics. This chapter concludes with three interesting instances of their use.

8.16 Note on a Class of Problems in Elementary Dynamics

Recent experience in teaching dynamics to first-year undergraduates suggests that a certain class of problems is being taught rather badly in schools. I refer to problems involving two or more particles when one is not required to find the values of the internal forces of the system, but merely asked to find, for example, the acceleration of one of the particles. There is a fairly widespread tendency to obtain the equation of motion of each separate particle of the system, thereby introducing a large number of unknowns which must be subsequently eliminated. But the required solution may often be obtained without the introduction of irrelevancies by considering the motion of the centre of gravity of the system, and by making use of the principle of conservation of energy.

As a representative example of the class of problems under discussion, consider the following worked example.†

'A smooth isosceles wedge of mass M is placed on a smooth table, and carries a small smooth pulley at its summit. Two masses, m_1 and m_2, are attached to the ends of a string which passes over the pulley. Show that the acceleration with which the string passes over the pulley is

$$\frac{m_1 - m_2}{m_1 + m_2} \cdot \frac{M + m_1 + m_2}{M + (m_1 + m_2)\sin^2\alpha} \cdot g \sin\alpha, \qquad (1)$$

where α is the angle at the base of the wedge.'

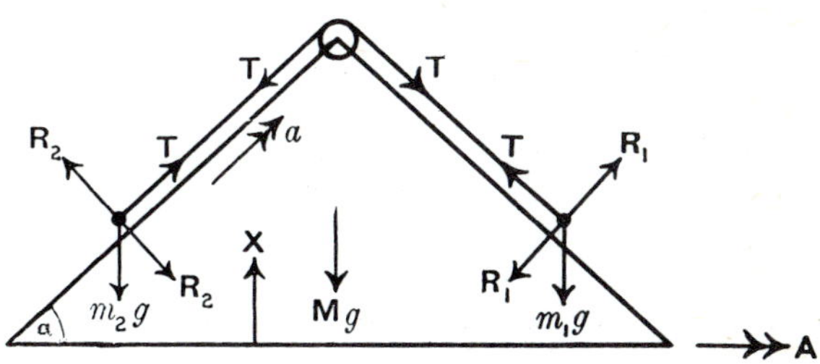

† Taken from p. 198 of *Mechanics* (Palmer and Snell).

The solution given by Palmer and Snell is outlined below.

Relative to the wedge, let m_1 and m_2 have acceleration a down and up the wedge respectively: and let A be the acceleration of the wedge relative to the table. Let T be the tension in the string, and R_1, R_2 the reaction between M and m_1, m_2 respectively. Let X be the reaction between the wedge and the table.

By considering the motion of m_1, m_2 and M separately, the following six equations are obtained:

$$\begin{aligned} T - m_2 g \sin\alpha &= m_2(a + A\cos\alpha), \\ m_2 g\cos\alpha - R_2 &= m_2 A\sin\alpha, \\ m_1 g\sin\alpha - T &= m_1(a + A\cos\alpha), \\ R_1 - m_1 g\cos\alpha &= m_1 A\sin\alpha, \\ R_2\sin\alpha - R_1\sin\alpha &= MA, \\ 2T\sin\alpha + (R_1 + R_2)\cos\alpha + Mg - X &= 0. \end{aligned}$$

Actually, only the first five equations are used; and by eliminating A, T, R_1 and R_2 from these equations, the required solution (1) is obtained.

The alternative solution below is not only simpler, but makes use of important dynamical principles.

Let y be the displacement of m_1 relative to the wedge, and let x be the displacement of the wedge relative to the table. Since the external forces have no horizontal component, it follows that if the system be initially at rest, then in the subsequent motion the centre of gravity of the system has no horizontal motion. This gives

$$Mx + (m_1 + m_2)(x + y\cos\alpha) = 0$$

during the whole motion, i.e.

$$x = -(m_1 + m_2)y\cos\alpha/(M + m_1 + m_2). \qquad (2)$$

The same relation is satisfied by the velocities $\dot{x}$, $\dot{y}$ and the accelerations, $\ddot{x}$ and $\ddot{y}$; this follows immediately by differentiating (2).

The principle of energy gives a further relation, viz. rate of increase of kinetic energy is equal to rate of loss of potential energy. This gives

$$\frac{d}{dt}\left[\tfrac{1}{2}M\dot{x}^2 + \tfrac{1}{2}(m_1 + m_2)(\dot{x}^2 + \dot{y}^2 + 2\dot{x}\dot{y}\cos\alpha)\right] = \frac{d}{dt}\left[(m_1 - m_2)gy\sin\alpha\right],$$

i.e.

$$(M+m_1+m_2)\dot{x}\ddot{x}+(m_1+m_2)[\dot{y}\ddot{y}+(\ddot{x}\dot{y}+\dot{x}\ddot{y})\cos\alpha] = (m_1-m_2)g\dot{y}\sin\alpha. \qquad (3)$$

If we use in equation (3) the relations between $\dot{x}, \dot{y}$ and between $\ddot{x}, \ddot{y}$ given by (2), we readily obtain $\ddot{y}$ in the form given by (1).

Our method depends effectively upon the solution of two equations in two unknowns rather than on five equations in five unknowns; and since the internal forces T, R_1, R_2 have not been introduced, the question of their elimination does not arise.

8.17 Problems on 'Variable Mass'

Newton when he had, as he said, 'the whole ocean of truth lying uncharted before him' succeeded in mapping extremely well when he stated his second law of motion in terms of rate of change of momentum. We find for example the electron fitting neatly into this formulation, and we can set our science pupils to integrate the uniform field equation:

$$\frac{d}{dt}\left\{\frac{m_0 v}{\sqrt{1-v^2/c^2}}\right\} = k$$

The success of the rate-of-change-of-momentum routine has led unfortunately to a number of unsound applications, which are often classed together as 'variable mass problems'. Two examples will be given, for both of which an unsound treatment has been invited by the wording of questions at *S*-level by one of our examining boards.

Example 1. A falling drop of water is gaining mass by aggregation from the surrounding medium according to a given law. Given the initial mass m_0, to find an expression for the velocity at time t.

The unsound short cut is to consider the drop a mechanical system and to equate its instantaneous weight to the rate of change of 'its' downward momentum. This gives the right result *provided the mass gained by accretion was previously at rest.*

A correct treatment must deal with a system of particles clearly defined, into which no new particles come during the discussion. For beginners, as well as for others who like a clear picture of what is going on, it seems advisable to consider a short interval between time

t and $t+\delta t$. The system consists of the water which makes up the complete drop at the later instant, the mass being $m+\delta m$, of which δm was picked up during the interval. The drop's velocity is v at time t and $v+\delta v$ at time $t+\delta t$. To emphasise the importance of correct principle the swept-up mass is supposed to be falling with velocity u.

We equate the increase in downward momentum of the system to the impulse of external force on it (and there is no awkwardness about force between m and δm because such forces are internal). We write:

$$(m+\delta m)(v+\delta v)-(\delta m \,.\, u+mv) = mg\delta t$$

and the differential equation follows. For solution we must know the law of aggregation, e.g.: $m = m_0+kt$.

Example 2. A rocket is rising vertically against negligible air resistance. Mass issues at constant rate k, at constant *relative* velocity U.

A correct treatment considers the rocket and its contents at time t and total mass m as the 'system'. After lapse of time this is in two parts,

(i) mass $(m-k\delta t)$ with upward velocity $v+\delta v$, and
(ii) mass $k\delta t$ with upward velocity $v-U$.

Equating the increase in upward momentum of the system to the impulse of the upward force, which need only be stated to the first order, we have

$$(m-k\delta t)(v+\delta v)+k\delta t(v-U)-mv = -mg\delta t,$$

which gives

$$m\frac{dv}{dt} = kU-mg$$

where $m = m_0-kt$.

The constant term kU is the rocket-propulsive force of the rocket engineer. In a recent examination paper this was introduced as a given constant force F. The pupil rocket-enthusiast who then wrote down $F-mg$ equal to $m\frac{dv}{dt}$ in this instance disagreed with examiner, whose printed result required the automatic use of $\frac{d}{dt}(mv)$. As in example 1, there are special conditions in which this is correct; in this case, that the gases after efflux at every stage of the motion have no momentum. It is clear therefore that if a correct equation is to be derived the student must be told the whole story and must then equate impulse to change of momentum for a correctly constituted system.

In conclusion, after pointing out some pitfalls in the paths of those who frame as well as of those who try to solve such problems one would like to challenge those who have read so far, to frame a question about an express train picking up water at speed (*a*) when the driver takes action to maintain his speed steady, and (*b*) when he does not. To state the assumptions so as to make the question soluble is no mean intellectual exercise.

8.18 Motion of a Rocket

The problem of determining the motion of a rocket is usually solved by a method involving the use of infinitesimals. The method, well known, is to write down the momentum of the rocket at time t and equate this to its momentum at time $t+\delta t$ together with the momentum of the additional burnt gases expelled in the interval δt.

The following method is an alternative approach based on the construction of a simple integral equation. Let

M_0 = initial mass of rocket at time $t = 0$,

$m(t)$ = mass burnt per unit time at time t,

$V(t)$ = efflux velocity of gases (relative to the rocket) at time t,

$v(t)$ = velocity of rocket at time t.

The mass of the rocket at time t is

$$M_0 - \int_0^t m\,dt,$$

and its momentum is then

$$\left\{M_0 - \int_0^t m\,dt\right\}v.$$

Now at time t, the momentum of the expelled gases (measured in the direction of motion of the rocket) increases at the rate $m(v-V)$. Hence the momentum of all the burnt gas at time t is

$$\int_0^t m(v-V)\,dt.$$

For a rocket moving in free space, the total momentum of the entire system is constant. Hence

$$\left\{M_0 - \int_0^t m\,dt\right\}v + \int_0^t m(v-V)\,dt = \text{constant.}$$

Differentiating this equation, we obtain

$$-mv + \left\{M_0 - \int_0^t m\,dt\right\}\frac{dv}{dt} + m(v-V) = 0,$$

so that

$$\frac{dv}{dt} = mV \Bigg/ \left\{M_0 - \int_0^t m\,dt\right\}.$$

REFERENCES

8.1. *Geometrical Proof of $v^2 = u^2 + 2fs$*
'H.V.S.' 1870, Vol. 30, No. 288, p. 36
8.2. *A First Lesson on Kinematics—Application of Graphs*
K. S. Snell 1526, Vol. 25, No. 265, p. 170
8.3. *A Walk in the Rain*
A. Sutcliffe 2721, Vol. 41, No. 338, p. 271
8.4. *Implicit Assumptions in Standard Problems*
M. A. Porter 1822, Vol. 29, No. 284, p. 79
R. V. H. Roseveare 2264, Vol. 36, No. 315, p. 44
8.5. *Some Remarks on the Coefficient of Restitution*
W. H. Osbourne 1235, Vol. 21, No. 243, p. 155
8.6. *An Elementary Collision Problem*
A. G. Walker 1188, Vol. 20, No. 238, p. 142
8.7. *Theorems on Centroids*
A. Lodge 1168, Vol. 19, No. 236, p. 367
8.8. *Further Centre of Gravity Problems*
R. E. Green C 79, Vol. 45, No. 354, p. 336
C. D. Langford C 69, Vol. 45, No. 352, p. 139
L. H. Le Bon 3017, Vol. 46, No. 356, p. 144
8.9. *Moments of Inertia*
W. Hope-Jones 2520, Vol. 39, No. 328, p. 145
D. Bolam & J. Wilkinson C 78, Vol. 45, No. 354, p. 335

8.10. *Proof of the Formula for Acceleration of a Point Moving with Uniform Speed in the Circumference of a Circle*
'R.F.M.' 988, Vol. 15, No. 212, p. 360
8.11. *The Elementary Theorems in Analytical Statics*
A. C. Heath 1193, Vol. 20, No. 238, p. 148
8.12. *The Reduction of the Mass-Accelerations and the Momentum of a Rigid Lamina*
K. S. Snell & A. H. G. Palmer 1195, Vol. 20, No. 238, p. 150
8.13. *A Geometric Side-line on 8.12*
'S.L.G.' 1750, Vol. 28, No. 282, p. 183
8.14. *A Problem on Friction*
R. E. Green C 55, Vol. 44, No. 349, p. 195
8.15. *Braking Efficiency in Motor Cars*
A. P. Rollett 1467, Vol. 24, No. 260, p. 212
8.16. *Note on a Class of Problems in Elementary Dynamics*
T. J. Willmore 2319, Vol. 37, No. 319, p. 44
8.17. *Problems on 'Variable Mass'*
A. J. Noakes 2680, Vol. 41, No. 336, p. 116
8.18. *Motion of a Rocket*
F. Chorlton 2363, Vol. 37, No. 322, p. 268

Index

acceleration 8.1, 8.10, 8.12, 8.16
angular momentum 8.12, 8.13
Apollonius, theorem of 5.3, 5.5
approximations 1.2, 1.3, 4.7, 4.19, 4.20
area 1.2, 4.2, 4.7, 5.11, 5.14
—under graph 4.1, 8.2
arithmetic mean 3.10, 3.11
average speed 1.1, 8.2

binomial theorem 3.12, 3.13

centre of gravity 7.1, 8.7, 8.8, 8.9, 8.12, 8.16
chords 2.13, 5.8, 5.15
circles 1.2, 2.13, 5.5, 5.6, 5.7, 5.8, 5.12, 6.5, 6.6, 6.7 ,8.14
—coaxial 3.1
circular functions 4.3, 4.6, 4.9, 4.13, 6.1, 6.2, 6.3, 6.4, 6.7, 6.8, 8.3
circular motion 4.9, 5.10, 8.10
coin tossing 2.5
collisions 8.5, 8.6
complex numbers 3.7, 5.8
components 7.2, 8.16
conics 2.12, 3.2, 3.3, 8.14
constant acceleration formulae 8.1
constants of integration 4.4
continuity 4.16
convergence 1.5, 3.11, 3.21, 3.23, 4.14, 4.15
cosine rule 5.5, 6.2, 6.4, 6.8
couples 8.11, 8.12
cube roots 1.3
cubic functions 3.5, 3.6, 3.7, 3.8, 3.16
cyclic quadrilaterals 5.4, 5.5, 5.7, 5.8
cyclical symmetry 2.7, 5.9

decimals 1.1
differences 2.10, 3.15, 4.3

differential equations 4.22, 4.23
directed numbers 1.1, 5.1
divisibility 1.1, 3.9, 4.18
division 1.1, 3.9, 4.18

elevation 6.3
energy 8.16
equations, theory of 3.1, 3.2, 3.3, 3.4, 3.5, 3.6, 3.7
expected value 2.2, 2.8
exponential function 4.6, 4.11, 4.12, 4.19

Fermat 1.3
Fibonacci numbers 3.13
formulae 1.4, 8.1
fractions 1.1, 1.2, 1.5
friction 8.4, 8.14, 8.15

generating function 2.3
geometric mean 3.10, 3.11
geometric progression 1.5, 4.14
geometrical constructions 3.1, 5.9, 5.10, 5.12, 5.13, 5.15
graphs 3.7, 3.8, 4.9, 8.2
great circles 6.6

harmonic series 4.15
Hooke's law 8.4
hyperbola 8.14
hyperbolic functions 4.6

impulse 8.5, 8.12, 8.17
interest 4.12
instantaneous centre 8.12, 8.13
invertible numbers 1.6
iteration 4.7, 4.18

knight tours 2.7

latitude and longitude 6.5, 6.6
limits 1.5, 3.11, 4.3, 4.10, 4.13, 4.16, 4.18, 4.19
line symmetry 5.1, 5.8, 5.9

linear momentum 8.4, 8.5, 8.12, 8.17, 8.18
logarithmic function 4.10, 4.11, 4.12, 4.20

Maclaurin series 3.21
matrices 3.19, 6.8
maximum and minimum values 3.10, 4.8, 8.3
mean value theorems 4.17
models 6.5, 6.7
moments
—of forces 8.11
—of inertia 8.9, 8.12, 8.13
momentum 8.4, 8.5, 8.12, 8.13, 8.17, 8.18

networks 2.7
Newton-Raphson method 4.7, 4.18
Newton's experimental law 8.5, 8.6
nomograms 3.3, 3.16
number scales 4.15

odds 2.1

paper folding 3.20
parabola 2.12, 3.2, 3.3
particular integrals 4.22
Pascal's triangle 3.13
periodic motion 8.4
permutations and combinations 2.4, 2.6, 2.9, 2.10, 3.13
planes 2.11
point symmetry 3.8, 5.9
polynomials 3.9
power series 3.12, 3.14, 3.15, 4.20
proportion 3.22, 4.3

quadratic functions 1.4, 3.1, 3.2, 3.3

radius of curvature 4.21
ratio 1.1, 1.5, 4.3, 5.6, 8.4
rational functions 3.17, 4.6
rectangular distribution 2.8

restitution 8.5, 8.6
Rolle's theorem 4.17
Roman numerals 1.1
rotation 8.9, 8.12, 8.13
rotational symmetry 5.1

scalar product 6.8, 7.1, 7.2
sequences 1.4
similarity 5.2, 5.3, 5.11, 5.15, 5.16
simple harmonic motion 5.10
Simpson's rule 4.19, 4.20
sine rule 6.2, 6.8
square and rectangle 1.2, 3.14, 3.20, 5.6, 6.1
square numbers 1.4
square roots 1.2, 4.19
symmetry 2.7, 3.8, 5.1, 5.8, 5.9
systems of forces 8.11, 8.14, 8.16

tangents 3.2, 3.7, 3.8, 4.9, 5.8, 5.12
tangrams 5.14
tessellations 5.9, 5.11
tetrahedron 2.6, 3.13
topology 2.7, 4.16
transformations 2.6, 4.9
triangles 5.3, 5.5, 5.10, 5.16, 5.17, 6.2, 6.4, 6.8, 7.1, 8.8, 8.9
—Heronian 5.5
—isosceles 5.5, 5.16
—right angled 5.2, 6.1, 6.3
triangular numbers 1.4, 3.13

unit vector 7.2

vector product 7.2
velocity 8.1, 8.2, 8.3, 8.4, 8.12

waiting times 2.8